MW00437292

Jump Linear Systems in Automatic Control

Jump Linear Systems in Automatic Control

MICHEL MARITON

MATRA SEP
Imagerie et Informatique
Saint-Quentin-en-Yvelines, France

MARCEL DEKKER, INC. New York and Basel

Library of Congress Cataloging in Publication Data

Mariton, M.
Jump linear systems in automatic control.
Includes bibliographical references.
1. Control theory. 2. Jump processes. 3. Linear systems. I. Title.
QA402.3.M3418 1990 629.8'312 90-2721
ISBN 0-8247-8200-3 (alk. paper)

This book is printed on acid-free paper.

Copyright © 1990 by Marcel Dekker, Inc. All Rights Reserved.

Neither this book nor any part may be reproduced or transmitted in any form or by any means, electronic or mechanical, including photocopying, microfilming, and recording, or by any information storage and retrieval system, without permission in writing from the publisher.

Marcel Dekker, Inc.
270 Madison Avenue, New York, New York 10016

Current printing (last digit):
10 9 8 7 6 5 4 3 2 1

PRINTED IN THE UNITED STATES OF AMERICA

Preface

Engineers frequently face problems which contain a genuine uncertainty concerning the current characteristics of the plant for which a device, filter or controller is to be designed. Adaptive solutions are one response to this difficulty, and provide on-line improvements to compensate initial ignorance and changes in the environment.

Pursuing this idea, adaptive systems have been attracting a great deal of research in the fields of control, communications and signal processing, as witnessed by numerous monographs, journals and conference proceedings. However, most of the attention up till now has been focused on the case where the parameters to be adapted evolve on a *continuum*, disregarding the fact that adaptation is triggered mainly by discrete events and often concerns parameters in a *discrete* set. The history of the system can be viewed as a succession of phases, each corresponding to a regime of operation, with sudden transitions that manifest themselves through jumps of the parameters of the model.

The distinctive feature of the systems studied in this book is the presence of this discrete parameter process which must be appended to the usual variables on a continuous space in order to describe the dynamics completely. To reflect this representation on discrete and continuous spaces, the name *hybrid* has been used.

Just as the designer has to make decisions under uncertainty, he often has to consider stochastic models where the future control trajectories and the present solution do not determine completely the future of the plant. The special class of stochastic systems considered here stems from applications where the transitions between the different regimes have to be considered as random; in other words, randomicity enters the plant primarily through its discrete component, and the natural model for the parameter process is given by a stochastic jump process.

As chapter 1 illustrates, hybrid systems are emerging as a convenient mathematical framework for the formulation of various design problems in fields such as target tracking, fault tolerant control and manufacturing processes. This monograph was conceived as an attempt to present a comprehensive body of theoretical results in order to confront these challenging applications. With this objective in mind, the pattern applied is as follows: chapter 2 gives the system theoretic properties of hybrid models. Optimal control law synthesis is studied in chapter 3, and chapter 4 analyses the robustness of the controlled system. In chapter 5 we refine our model with additive and multiplicative noises, which leads to optimal filtering problems dealt with in chapter 6. Most realistic is the case where there is uncertainty on the current regime of operation. Reconfiguration (or discrete adaptation) then requires special care, and new solutions are described in chapter 7. Finally, in chapter 8 we consider more general models and indicate several open areas of research.

The book is primarily a research monograph and most of its results are recent. Given the remarkable modelling potential of hybrid systems, it should be appreciated by engineers in the areas of control and signal processing. It is also hoped that the book will contribute to attracting researchers into a field which is not only very rich theoretically but also forward-moving in its increasing number of applications. A graduate class in control could also use the material over a semester to illustrate concepts of stochastic systems theory. Appendices 1 and 2 would then play a key role in recalling the fundamentals of probability and control theories.

I would like to acknowledge my gratitude to several persons and institutions whose support made this work possible. First I would like to thank Professor Dave Sworder of the University of California at San Diego. Not only does his own research contribute to significant portions of this work, but he also provided a stimulating environment for the writing of the first manuscript. Without his friendly hospitality, this project would not have been accomplished. The support and guidance of Dr. Pierre Bertrand, who has always been ready to help with the important decisions over the years, is also gratefully acknowledged. My friends and colleagues at the Laboratoire des Signaux et Systèmes, Drs. Eric Walter and Luc Pronzato, read the manuscript and made many valuable suggestions, as did also Ph.D. student Yang Chun, whose thesis contributed to the material studied. I would also like to acknowledge the CNRS, the French national center for scientific research, which provides unique working conditions for young researchers, conditions from which I benefited during my association with the Laboratoire des Signaux et Systèmes. The Direction des Recherches Etudes et Techniques supplied additional support for my stay at La Jolla (contract DRET no.86/1391). Finally I would like to thank MATRA and MS2i, in particular Mr. Jean Broquet who believed in my projects and Dr. G. Ruckebusch. The manuscript has been expertly typed and corrected by Mesdames Karen Leterte and Eve Salinas. Finally, I would like to mention the help from Marcel Dekker, Inc., especially Ms. Ruth Dawe and Ms. Beth Wooster.

Michel Mariton

Contents

1

Hybrid Dynamic Models

The purpose of this introductory chapter is to present the basic model that we shall study and to settle corresponding vocabulary and notational conventions. Using examples, we insist on the modelling aspect and on the process of abstracting the main features of a given application into a hybrid model.

1.1 An Example: Target Tracking

In recent years concern with the safety of air traffic near crowded airports and tragic accidents have prompted research into more sophisticated radar tracking algorithms which could help air controllers monitor incoming and outgoing aircraft. A specific difficulty appears in areas like California where there exists heavy traffic of small highly maneuverable private aircraft interfering with commercial jet liners.

In a military context, a related problem is the so-called evasive target tracking problem where it is desired to keep track of an object which is maneuvering quickly in an attempt to evade its pursuer. The performance of the tracking system heavily depends on the accuracy and sophistication of the model used to describe the target dynamics.

A first element of the model is based on the equations of motion, relating variables like horizontal position, heading speed, bank angle or flight path angle. Physics provides a set of differential equations in $\mathbf{R}^n$, where n is the number of variables we retain.

A well-known problem when multiple targets are to be tracked is to associate radar data to the various tracks, which might be difficult when a clutter of signals in a given region exists. However, another less usual phenomenon is of primary interest here, namely the effects of sudden changes in the acceleration of the target. The trajectory can be divided into several sequences when the aircraft flies with (almost) constant acceleration, bank angle and flight path angle; regimes of flight typically considered are ascending flight after take-off, turning, accelerated flight, uniform cruise motion or descending approach flight. The transitions between these regimes are discrete and depend primarily on the pilot decisions which are in turn influenced by weather data, mission controller indications and on-board information such as fuel consumption, etc. or perception of the threat. Depending on the current regime of flight chosen by the pilot, the coefficients of the dynamic model have to be adjusted.

Of course evasive maneuvers are chosen to confuse the pursuer, and are therefore characterized by frequent, large, irregular and seemingly irrational acceleration changes. Figure 1.1 shows a typical realization where sudden acceleration is induced by the occurrence of a point process.

From the tracker point of view, these transitions are perceived as random, and a model which makes the least use possible of other a priori information is a hybrid

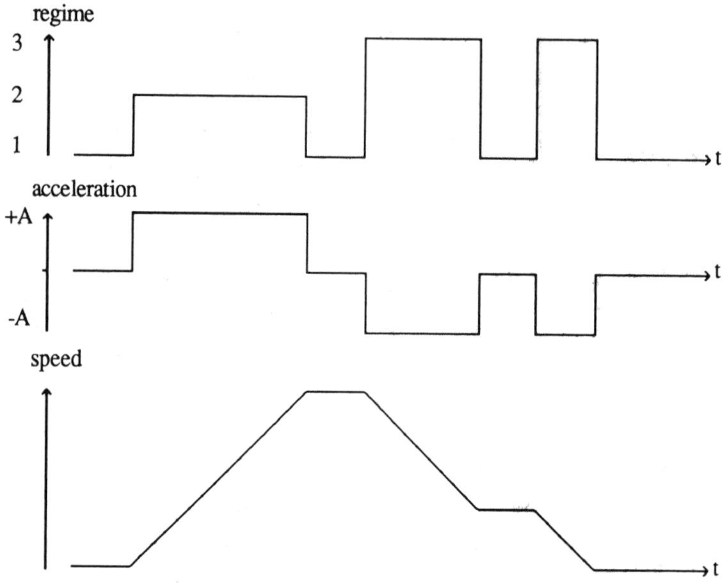

Figure 1.1 - Typical acceleration changes.

stochastic model with continuous dynamics perturbed by random transitions of a regime variable.

Our discussion so far has been confined to conventional point tracking. However, tracking sensor suites, mostly in the military domain at present, include sensors with imaging capacity (e.g. IR images). Clearly the information contained in the image is valuable for the tracking task: the image processor can tell whether the target is in or out of the focal plane or can even classify the type of target encountered. This information is well described by a discrete variable, indexing words from a given alphabet, for example "tank in the focal plane" or "jeep of type T" etc. This declaration is polluted by classification errors and the accumulation of processed images (frames) is needed to reduce uncertainty through a temporal processing. Figure 1.2 illustrates this idea for a weapon platform.

Again sudden changes of the target orientation or of the engagement scenario are naturally described as random transition of the discrete variable and these impact on the target location continuous variables through a change of dynamics.

1.2 The Basic Model

We consider hybrid dynamics with a euclidean part and a discrete part. For the euclidean part of the process dynamics, we shall note $x_{pt} \in \mathbf{R}^n$ the *plant state* and restrict ourselves to finite dimensional systems. A control (or input) vector $u_{pt} \in \mathbf{R}^m$ and an observation (or output) vector $y_{pt} \in \mathbf{R}^p$ are used to represent the action variables (u_p)

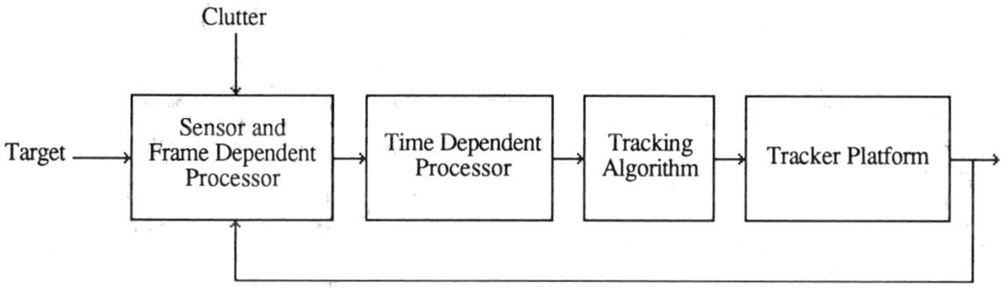

Figure 1.2 - Tracking with an imaging sensor.

and the available measurements (y_p). In the target tracking problem x_{pt} contains target location variables (position and velocity), u_{pt} the tracker platform orientation command and y_{pt} sensor readings (range, azimuth angles, etc.). The discrete variable $r_t \in \mathbf{S} = \{1, 2, ..., N\}$, called the *plant regime* models, for example, the presence of a maneuvre, target classification (friend or foe?), etc.

Most of the rest of the book is thus concerned with the discrete variable $r_t \in \mathbf{S}$, and three variables in euclidean spaces $x_{pt} \in \mathbf{R}^n$, $u_{pt} \in \mathbf{R}^m$, $y_{pt} \in \mathbf{R}^p$ and their interactions.

A mathematical model must of course be introduced to describe the coupled dynamics of these variables. However, it is appropriate to pause for a moment here to stress further the meaning of the adjectives hybrid, discrete and continuous in the present context. As the above example clearly shows, hybrid refers to the state-space on which the systems studied are living: it is said to be hybrid because it is the product of a euclidean vector space ($\mathbf{R}^n$) and of a discrete finite set ($\mathbf{S}$). This has nothing to do with the way time is measured in the system, and indeed hybrid models can be considered in continuous time or discrete time as well. We shall work in continuous time, and we give references in §1.4 for the transposition to discrete time.

Let $(\Omega, \Xi, \mathcal{P})$ be a probability space with an associated nondecreasing family of σ-algebras $\Xi_t \subset \Xi$. The euclidean variables x_{pt}, u_{pt} and y_{pt} obey a piecewise deterministic model

$$\begin{aligned} \dot{x}_{pt} &= f(x_{pt}, u_{pt}, r_t, t) \\ y_{pt} &= h(x_{pt}, r_t, t) \end{aligned} \tag{1.2.1}$$

where f and h are analytic mappings from $\mathbf{R}^n \times \mathbf{R}^m \times \mathbf{S}$ into $\mathbf{R}^n$ and from $\mathbf{R}^n \times \mathbf{S}$ into $\mathbf{R}^p$, respectively. The distinctive feature of (1.2.1) is the dependence of the dynamics and observation equations on the regime. We are interested in situations where the regime transitions are random (evasive manoeuvres, etc.). To describe the jump dynamics we introduce the indicator of the regime $\phi_t \in \mathbf{R}^N$ (N = card $\mathbf{S}$) with components $\phi_{ti} = 1$ when $r_t = i$ and $\phi_{ti} = 0$ otherwise, i = 1 to N. The indicator is piecewise constant and we shall most often work with the simplest model for the regime transitions, a Markov chain given as

$$d\phi_t = \Pi'\phi_t dt + dM_t \tag{1.2.2}$$

with M_t an Ξ_t - martingale and Π the chain generator, an N x N matrix. The entries π_{ij}, i, j = 1, N of Π are interpreted as transition rates

$$\mathscr{P}\{r_{t+\Delta} = j \mid r_t = i\} = \begin{cases} \pi_{ij}\Delta + o(\Delta) & i \neq j \\ 1 + \pi_{ii}\Delta + o(\Delta) & i = j \end{cases} \tag{1.2.3}$$

Observe that the total probability axiom imposes $\pi_{ii} = -\sum_{\substack{j=1 \\ j\neq i}}^{N} \pi_{ij}$ and π_{ii} negative.

In many situations the piecewise deterministic state model (1.2.1) is completed with the influence of wide-band noises, e.g. the addition of radar clutter to the target location measurement in the tracking scenario just described. This is considered later in the book, starting with chapter 5. The important point we shall use in what follows is the Markov property of the (state, regime) pair.

Indexed by r_t, (1.2.1) in fact represents a family of N dynamic and observation equations, f(-, -, i, -) and h(-, i, -), i = 1 to N. This is used in the tracking application to describe changes in positions and velocity time constants when the target manoeuvres or changes in the IR signature output as the target orientation is modified.

Because of the difficulties inherent in the analysis of nonlinear dynamics, most attention is given to the special case where f and h are linear and stationary

$$\begin{aligned} \dot{x}_t &= A(r_t)\, x_t + B(r_t)\, u_t \\ y_t &= C(r_t)\, x_t \end{aligned} \tag{1.2.4}$$

with matrices A, B and C of corresponding dimensions. Very often we denote the current regime by an index (e.g. A_i stands for $A(r_t)$ when r_t = i) and, more generally, we use either r_t, ϕ_t or i to designate the regime. For example, $B(r_t)$, $B(\phi_t)$ or $\sum_{i=1}^{N} B_i\, \phi_{ti}$ are three possible expressions for the value of B at time t and we shall choose one or another indifferently as is most convenient. From the context there should be no confusion and we shall make no further reference to this simplifying abuse of notation. We shall also often use the convention $[A_i, B_i, C_i, i = 1, N, \Pi]$ to summarize the model (1.2.2), (1.2.4).

Although A, B and C do not depend explicitly on time most of our results would hold for "non-stationary" jump models. Similarly the transition rates can be made time-dependent $\Pi = \Pi(t)$. As long as t corresponds to the absolute time the markov property is preserved, but if time is counted relative to the last transition (sojourn-time dependent rates) the situation is more involved and our solutions must be refined. One of these non markov situations is studied in chapter 8.

Going from (1.2.1) to (1.2.4) involves linearization around some nominal set points, which is justified in many practical situations but nevertheless requires special attention in the present context where the state space is hybrid. Indeed, the linearization procedure underlying an approximation like (1.2.4) is worth a detailed discussion and this will be done at several points throughout the book when the validity of (1.2.4) or the consequences of linearization significantly affects the discussion (see in particular chapters 4 and 7). Note that to keep track of the linearization, variables x_p, u_p and y_p have been replaced in (1.2.4) by $x \in \mathbf{R}^n$, $u \in \mathbf{R}^m$ and $y \in \mathbf{R}^p$ which stand for the deviations of x_p, u_p and y_p with respect to some nominal level.

The jump linear model (1.2.2), (1.2.4) is the basic model to be studied here and will be used as a starting point to build more sophisticated models, for example by adding noises to (1.2.4) or by considering non markovian transitions instead of (1.2.2). This model is simple enough to preserve mathematical tractability while abstracting the main features of significant applications, e.g. the target tracking example of §1.1 or the three examples described below.

1.3 Three Other Applications

We consider other typical situations where hybrid jump models arise naturally: a manufacturing process, a solar thermal receiver and fault-tolerant control systems. In each case it is explained how the random environment manifests itself through sudden changes in the parameters of the model leading to a stochastic interaction between discrete and continuous variables and multiple regime behaviors. These examples, together with the target tracking problem, will be used again later in the book either for illustrative purposes or to motivate some refinements of the model.

1.3.1 A manufacturing process

To generate benefits and increase market shares, companies must produce with a high flexibility to follow the versatile customer demand and the question of the desirable level of inventory must be resolved. With the advent of cheap and reliable microcomputers, considerable hopes were placed during the seventies in a decentralized/hierarchical production management approach where each elementary production unit would be controlled locally with some coordination provided at a supervisory level. Many companies started reshaping their factories or even building new ones to accommodate the computers and robots that were expected to be the panacea for increased productivity. They quickly realized, however, that it was not so easy to make these machines work together and the confusion increased to the point where, paradoxically, automation was thought of as a source of trouble rather than improvement (machine failures, unprepared workers, etc.).

A basic explanation is that the computer aspect of the problem was overemphasized to the detriment of the control aspect so that a global perception of the behavior of an automated plant was certainly missing, and, especially, the dynamics of systems integrating different components were poorly understood.

A model of the manufacturing process can be considered as one of the desirable assets to conduct a rational analysis since, even if such a model is bound to be a simplified approximation of reality, conclusions based on its qualitative behavior can be extrapolated with reasonable confidence to real industrial processes.

Failure-prone manufacturing systems are intrinsically hybrid, i.e. they evolve on a product space $\mathbf{R}^n \times \mathbf{N} \times \mathbf{S}$ where $\mathbf{R}^n$ and $\mathbf{S} = \{1, 2, ..., N\}$ are as before and $\mathbf{N}$ is the set of integers.

The $\mathbf{N}$ component is well-suited to the description of the number of products manufactured, while on $\mathbf{R}^n$ one might consider variables such as machine power consumption or the storage of raw material. The finite set $\mathbf{S}$ is a list of possible failures with the simplest situation being a single machine system where $N = 2$ and a working (1)/failed (2) classification. Figure 1.3 gives a block diagram of a production system described in this way.

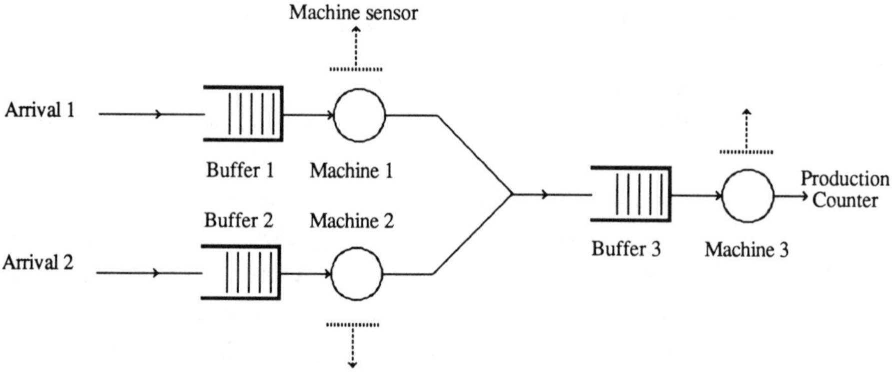

Figure 1.3 - A simple production system.

An important point for improving performance is the quick and reliable detection of degradations to prevent full failures and stoppages in production. The detector may use hybrid measurements on $\mathbf{R}^n \times \mathbf{N}$. The integer part of the observations is delivered by counting processes recording the number of products manufactured over time. A low delivery rate is, for example, considered as indicative of a possible incident along the production line. A quality test is also performed on the products, classifying them as satisfactory or defective. Again a high rate of defective products may signify a failure. However, these quality tests are not perfect and may generate errors such as counting a good part as defective. If the output rate is low (long production line, etc.) a significant delay is also incurred before the product counting processes start to reflect an incident. Fortunately, a secondary set of measurements is often available, which is best described by the continuous part of the model on $\mathbf{R}^n$. Depending on the process under consideration, existing manufacturing systems are equipped with sensors monitoring quantities like machine power consumption, vibration levels and raw material storage fluctuations. Usually the corresponding data are simply displayed in the control room for the human operator, who monitors recordings and uses expert knowledge about events such as drifts in power consumption when suspecting a possible incident. Alternatively, this situation may be abstracted in a mathematical model such as that of §1.2 and optimal filtering and detection theory can be invoked to design an automatic plant monitor. Chapter 6 studies some possible filters.

1.3.2 A solar thermal receiver

To produce electrical power directly from the sun's energy, a large solar electrical generating system can be designed with a field of movable mirrors, called heliostats. These are used to focus the energy of the sun on a central boiler where the incoming water flowing through a panel is transformed into superheated steam, which in turn drives a conventional turbine/generator pair. Solar electrical generating systems have been constructed in the Californian desert at Daggett, and at Font-Romeu in the French Pyrenees, following the oil crisis of the early seventies. To go beyond these experimental studies and produce power on an industrial basis, one of the key control challenges is the achievement of an accurate regulation of the steam temperature in the boiler. This temperature is the main performance parameter and should be kept close to nominal value, but at the same time the feedwater flow rate must prevent the metal temperature from getting too high. With current technology, these requirements are further narrowed down by the fact that the solar energy panel will warp in a matter of minutes when it becomes overheated.

The designer is thus faced with a difficult problem where demanding requirements are to be met as well as stringent security constraints, and it is clear that an accurate model of the receiver dynamics is needed.

Some models which have been derived to evaluate control strategies basically relate physical variables like metal temperature, outlet pressure, input water enthalpy, etc. through nonlinear thermodynamic relations. For a given level of insolation we can assign values to the coefficients of such a model which are typically heat transfer coefficients, metal mass, surface areas, and so forth, and constitute the continuous part of the hybrid model. However, there is another less usual phenomenon which must be described: the motion of clouds over the field of heliostats. Even if the facility is built in a sunny area like California, there will be partly cloudy days during which clouds briefly cover the heliostats before moving away. It has been observed that this produces a variation of the insolation at the central boiler of up to 80%, thus heavily influencing the thermodynamic nonlinear equations. For example, the responsiveness of the metal temperature to a modification of the water flow rate is directly related to the insolation level, with a fast response at high levels and a more sluggish one when the sun is obscured by cloud. These sudden changes in insolation are essentially unpredictable locally and manifest themselves in a discrete and random fashion. To make the model as independent as

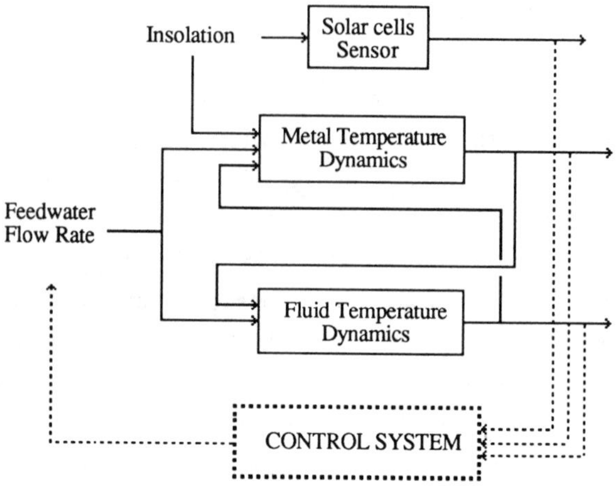

Figure 1.4 - Schematic diagram of the boiler.

possible of external sources of information, it thus seems natural to introduce a discrete jump variable, the regime. It describes the motion of clouds over the plant and we make the thermodynamic equations on the continuous space contingent on this discrete variable to describe the way a random jump in the insolation level affects the behavior of temperatures, pressures, flows and so on (see figure 1.4).

In order to keep the dimension of the global model small enough, a quantization of the insolation level is introduced, for example 20% and 80% of the full insulation for a large dense cloud and a small cloud respectively. An approximation is thus introduced, but it is outweighed by the mathematical tractability resulting from limiting the regime variable to a finite set. The motion of clouds being random, the transitions of the insolation from one quantized level to another are expressed as jump probabilities where the rates are now estimated from records of past data under typical weather conditions. This is done by matching the sample path behavior, essentially characterized by the mean time between two clouds, to a simple probabilistic mechanism.

The final model is thus very similar to the previous target tracking or manufacturing process models, with the characteristic interaction between a variable on a continuous space for the usual plant dynamics and a discrete random process for events such as manoeuvres or clouds which affect the system evolution. This fits nicely into our basic model (1.2.1), (1.2.2).

1.3.3 Fault-tolerant control systems

Modern technological systems rely on ever more sophisticated control functions to meet increased performance requirements: aircraft with negative centering is a well-known example, trading reduced stability for greater maneuvering capacities. The system being open-loop unstable, a vital stabilization function is performed by the flight control system, and it is then clear that the aircraft is highly vulnerable to incidents like the failure of components (actuators, sensors or on-board computers).

Large flexible structures for future space stations are another striking example. Solar arrays, antennas or mirrors of next generation spacecraft will be monitored and controlled by hundreds of sensors and actuators spread over their surface and, given the costs involved, mission times will be very long (e.g. twenty years for the NASA microwave radiometer). Maintenance will be difficult for these in-orbit systems and, with the large number of components involved, it is not an over-statement to say that the usual regime of operation will be a failed one, that is, a regime where some components have experienced failures.

In order to increase the reliability in the presence of such vulnerability, the flight control system will have to provide some kind of fault tolerance. Since Von

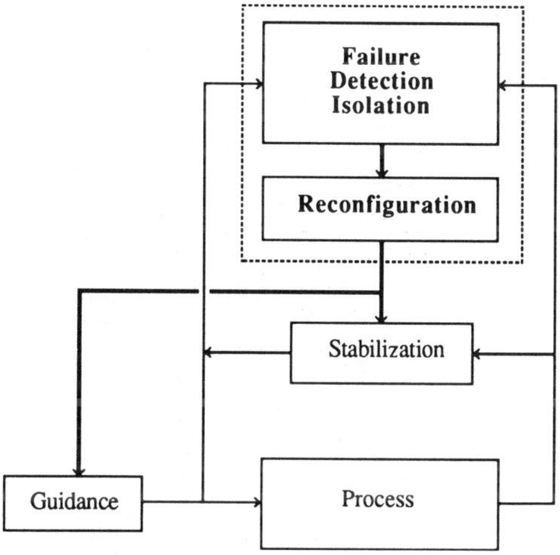

Figure 1.5 - A reliable control system.

Neuman, we know that redundancy is the basic ingredient for building a reliable system. This idea has been exploited in circuit theory to produce the so-called "no-down-time computers", in fact computers with triplex or quadruplex processing units and increased reliability.

For the past fifteen years fault-tolerant control systems have attracted a significant research effort and the basic structure of a reliable control system, with redundant actuators and sensors, which is now agreed upon is presented in figure 1.5.

The redundancy management system performs detection and isolation of the failures and reconfiguration of the remaining components and of the control algorithms.

To analyze the behavior of such a system we clearly need a description of the occurrence of failures and their influence on the process.

Consider, for example, a duplex system where two redundant controllers, C_1 and C_2, are used in parallel to control the plant P.

Excluding partial failures of a component, four regimes of operation, can be associated to figure 1.6 depending on which controller has failed:

regime 1	C_1	C_2
regime 2	$\not{C}_1$	C_2
regime 3	C_1	$\not{C}_2$
regime 4	$\not{C}_1$	$\not{C}_2$

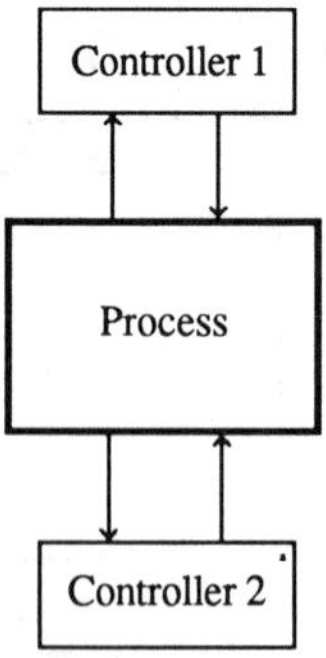

Figure 1.6 - A duplex control system.

Failures thus appear as discrete events that cause a transition, or a jump, of the regime. These events being random, they are characterized by transition probabilities and, noting p the probability vector in $\mathbf{R}^4$ with components

$$p_{it} = \mathscr{P}\{\text{regime} = i \text{ at time } t\}\ , \quad i = 1 \text{ to } 4$$

a classical model in reliability theory is given as

$$\dot{p}_t = \begin{bmatrix} -2\lambda & \lambda & \lambda & 0 \\ 0 & -\lambda & 0 & \lambda \\ 0 & 0 & -\lambda & \lambda \\ 0 & 0 & 0 & 0 \end{bmatrix} p_t \qquad (1.3.1)$$

where λ is the individual failure rate of C_1 and C_2 and the matrix in (1.3.1), often noted Π, is precisely the matrix of transition rates of the regime Markov chain of our basic model (1.2.2). In writing (1.3.1), note that a simultaneous failure of C_1 and C_2 was excluded (the transition from 1 to 4 has zero probability) which corresponds to the realistic situation where failures are rare events and the occurrence of a simultaneous failure highly unlikely. However, it would be easy to include this event in the hybrid model by replacing the zero in the first row of the matrix Π by a small positive coefficient. The possibility of maintenance is also conveniently described: for μ the individual maintenance rate of C_1 and C_2, the matrix of transition rates becomes

$$\Pi = \begin{bmatrix} -2\lambda & \lambda & \lambda & 0 \\ \mu & -(\lambda+\mu) & 0 & \lambda \\ \mu & 0 & -(\lambda+\mu) & \lambda \\ 0 & \mu & \mu & -2\mu \end{bmatrix}$$

The failures manifest themselves through a modification of the actuator-process-sensors cascade in figure 1.5. Typically the failure of a sensor introduces a bias or a drift in one of the measurement variables; similarly a failed actuator might produce a constant action on the process regardless of the command signal it receives. A physical fault in some parts of the plant also modifies the dynamics.

It is thus seen that fault-tolerant control systems are naturally described in terms of the interaction between a discrete random jump variable accounting for the occurrence of failures, and the more usual variables in a continuous space (e.g. attitude angles and their rates for the above aerospace examples), representing the plant dynamics.

1.4 Notes and References

We have introduced the notion of hybrid dynamics, characterized by a product space $\mathbf{R}^n \times \mathbf{S}$. A typical modelling for the plant state on $\mathbf{R}^n$ includes

- positions and velocities (target tracking)
- vibration level or power consumption (manufacturing system)
- steam pressure and metal temperature (solar receiver)
- aircraft attitude angles (fault-tolerant flight control system)

and for the plant regime in **S**

- maneuvres, orientations, classifications (target tracking)
- nominal, degraded or failed status (manufacturing and fault-tolerant systems)
- clouds crossing the field of heliostats (solar receiver)

Through equations (1.2.1) and (1.2.4) the regime indexes the state dynamics to describe the influence of maneuvres on target motion, the change of the operating point when a cloud obscurs the sun or the degraded behavior of a system with some failed components.

Let us mention briefly other situations where hybrid models are relevant. In the analysis of electrical power networks one usually deals with continuous variables like voltages or currents, but discrete events, typically lightning-induced line faults or modifications of the network topology also have a significant impact on the dynamic behavior of the network. It has even been conjectured that the major power shut-downs in Europe and the United States during the late seventies were due to an uncontrolled cascade of discrete events, leading the nonlinear continuous dynamics of interconnected networks into unstable regions.

The field of signal processing is also rich in situations where a hybrid description is natural: integrated communication networks are used to transmit, on a single medium, different types of traffic, such as voice, images and data, and the operator needs to identify on-line the type of traffic which is transmitted, for example to monitor the network load, or analyse trends in the users' habits. His task is therefore to extract from a continuous variable - the signal, a discrete variable - the type of signal transmitted. Moreover, the description of traffic generation commonly accepted fits exactly into the random jump mechanism, and a stochastic hybrid model is suitable to formulate the detection problem.

Econometrics can also benefit from the use of hybrid models. In a difficult environment, it seems reasonable to look for planning policies which, through the instruments which the government may use, could regulate key economic variables like inflation rate, unemployment level and household average income. To analyze such policies, a macroeconomic model would be useful. Because of the complexity of modern nations, economists restrict their analysis to categorical averages, and have derived aggregate models that relate continuous fluctuations of, typically, national income, consumption level, etc. However, this continuous evolution is influenced by other variables, such as the interest rate or the price of oil, which act as exogenous disturbances. On the time-scale of the averaged dynamics, the variations of these disturbances look much like random discrete processes with, for example, sudden jumps of the interest rates from 8.5% to 8.0% and back to 8.3% or sudden jumps of the price of a barrel of oil from $25 to $16. A simple model like this can of course only be conceived as an approximation, but as it reflects qualitatively the key interactions between discrete and continuous variables, it provides an attractive alternative to purely empirical policy assessment.

Future work on hybrid systems will certainly be influenced by the demands from the large field of applications in BM/C^3 (Battle Management/Communication, Command and Control). Analyzing the requirements of the SDI (Strategic Defense Initiative) as an example, it has been recognized in (Athans, 1987) that hybrid dynamic systems form the basic modelling paradigm for these problems.

We could continue this listing further, finding situations in image processing and local area networks where hybrid models are also relevant, but it is now more fruitful to turn to the mathematical abstraction outlined in §1.2 which captures the essential features of all these applications.

Hybrid linear models like (1.2.2) and (1.2.4) first appeared in the literature in the early sixties in (Krasovskii and Lidskii, 1961) and (Florentin, 1961). These pioneering papers were concerned with analysis of the linear case and did not study any particular applications. In the late sixties control problems for linear systems with abrupt changes in parameter values were considered by (Sworder, 1969). This was the beginning of a continued research effort which might be summarized by the surveys (Sworder, 1976, Sworder and Chou, 1985). It is beyond the scope of this introductory chapter to describe the width and depth of Sworder's contributions to the theory of jump linear systems, and numerous references to the multiple facets of his work will be made later in the book where they will be closely related to the material presented.

While (Sworder, 1969) attacked the optimal control problem from a stochastic maximum principle point of view, (Wonham, 1971) showed that a stochastic extension of dynamic programming can also be used. In this book a matrix version of the maximum principle and stochastic dynamic programming will both be employed, depending on which is more convenient for a given optimization problem (appendix 2 contains a brief exposition of the required material on stochastic control).

Regarding the application of jump models to fault-tolerant control systems, it seems that the idea first appeared in (Siljak, 1978), and (Birdwell, 1978) considered the design of reconfigurable feedback loops for the discrete-time case. Mention of this work also provides an occasion for stressing the fact that most of the material which will be presented in continuous time has an obvious analog in the discrete-time setting. Due to the simplicity of the transposition, the decision was made to avoid duplicating the presentation, but the reader more interested in discrete-time applications may find it helpful to study references such as (Birdwell and Athans, 1977, Birdwell, Castanon and Athans, 1978, Chizeck, 1982, Chizeck, Willsky and Castanon, 1986) or (Millnert, 1982). The more general study of jump linear systems in discrete time was again initiated by Sworder (Sworder, 1967, Blair and Sworder, 1975b). The idea of using redundancy to improve reliability is much older and its systematic analysis for circuit design is due to (Von Neuman, 1956, Moore and Shannon, 1956). In continuous time (Siljak, 1980, 1981) introduced jump linear models to analyze the stability of a multiplex control system, a path which has been followed by (Ladde and Siljak, 1983, Mariton and Bertrand, 1986) while (Montgomery, 1983) studied in this framework the placement of failure-prone actuators on large space structures.

The tracking application is detailed in the monograph (Bar Shalom and Fortman, 1988), building on (Bar Shalom, 1978, Bar Shalom and Birmiwal, 1982, Chang and Bar Shalom, 1987). The hybrid setting for target manoeuvres was also considered in (Moose, 1975) and Blom, in a series of papers (e.g. Blom, 1984a, b, 1986, 1987), proposed the Interacting Multiple Model solution which provides an efficient implementable filter (see also (Blom and Bar Shalom, 1988)). Regarding tracking with imaging sensors, the use of hybrid models like (1.2.1), (1.2.2) was initiated by Sworder, see (Haaland, Hutchins and Sworder, 1987) or (Sworder, 1987). Control problems for manufacturing systems were formulated with hybrid models in (Akella and Kumar, 1986, Bielecki and Kumar, 1986) and the solar thermal receiver application of §1.3.2 is taken from (Sworder, 1982, Sworder and Rogers, 1983, Sworder, 1984).

The rich modelling potential of hybrid systems with stochastic jumps is further illustrated by the analysis of transient electrical power networks (Willsky and Levy, 1978, Malhame and Chong, 1985) or their application to economic policy planning (Kazangey and Sworder, 1971, Blair and Sworder, 1975a, building on Samuelson, 1939)).

The ideas and vocabulary of one or the other of the above applications may pervade our presentation in the following chapters, but it is our ambition in this book to develop general theoretical tools which will, we hope, encompass problems encountered in the extensive field of hybrid control and signal processing applications.

A distinctive feature of the situations we have in mind is that they require some form of adaptation with respect to a discrete parameter. This is clearly reflected in the jump variable in (1.2.1), (1.2.2) and, in the following chapters, the adaptation mechanism is based on the plant regime or, if not available, on an estimate of it.

Due to the random transitions of the discrete variable, it appears that this work is an addition to the literature on stochastic adaptive systems. However, the original purpose here is to restrict attention to a subclass of problems with a specific jump model where a complete and rather refined theory can be constructed with a promising potential for applications. Related but more general points of view are found in (Caines and Chen, 1985) or (Hijab, 1987) which provide a relevant background to our study.

We have now introduced the basic mathematical objects we shall be dealing with and, given the information generated by the observations, our program now is to design filters and controllers to achieve adaptive control of our hybrid plant.

To a large extent this program has determined the plan followed in the forthcoming chapters: chapter 2 provides system theoretic foundations to the basic design methods studied in chapter 3 and robustness questions are raised in chapter 4. Attention is then focused on the influence of noise on hybrid models, going to increasingly general situations from chapters 5 to 7, finally culminating in a design which provides a fast and robust adaptation to regime jumps. The book concludes with chapter 8 in which extensions to a larger class of models are considered, together with open research directions.

References

Akella, R., and Kumar, P.R. (1986). Optimal control of production rate in a failure prone manufacturing system, IEEE Trans. Aut. Control, AC-31: 116.

Athans, M. (1987). Command and control (C2) theory: a challenge to control science, IEEE Trans. Aut. Control, AC-32: 286.

Bar Shalom, Y., and Fortman, T.E. (1988). Tracking and Data Association, Academic Press, New York.

Bar Shalom, Y. (1978). Tracking methods in a multitarget environment, IEEE Trans. Aut. Control, AC-23: 618.

Bar Shalom, Y., and Birmiwal, K. (1982). Variable dimension for maneuvering target tracking, IEEE Trans. Aero. Electr. Systems, AES-18: 621.

Bielecki, T., and Kumar, P.R. (1986). Necessary and sufficient conditions for a zero inventory policy to be optimal in an unreliable manufacturing system, Proc. 25th IEEE Conf. Decision Control, Athens, pp.248-250.

Birdwell, J.D., and Athans, M. (1977). On the relationship between reliability and linear quadratic optimal control, Proc. 16th IEEE Conf. Decision Control, New Orleans, pp.129-134.

Birdwell, J.D. (1978). On reliable control systems design, Ph.D. Dissertation, Elect. Systems Lab., Mass. Inst. Technology, report no. ESL-TH-821.

Birdwell, J.D., Castanon, D.A., and Athans, M. (1978). On reliable system design with and without reconfiguration, Proc. 17th IEEE Conf. Decision Control, San Diego, pp.419-426.

Blair, W.P. Jr., and Sworder, D.D. (1975a). Continuous-time regulation of a class of econometric models, IEEE Trans. Systems Man Cyber., SMC-5: 341.

Blair, W.P. Jr., and Sworder, D.D. (1975b). Feedback control of a class of linear discrete-time systems with jump parameters and quadratic cost criteria, Int. J. Control, 21: 833.

Blom; H.A.P. (1984a). A sophisticated tracking algorithm for ATC surveillance radar data, Proc. Int. Conf. Radar, Paris, pp.393-398.

Blom, H.A.P. (1984b). An efficient filter for abruptly changing systems, Proc. 23rd IEEE Conf. Decision Control, Las Vegas, pp.656-658.

Blom, H.A.P. (1986). Overlooked potential of systems with markovian coefficients, Proc. 25th IEEE Conf. Decision Control, Athens, pp.1758-1764.

Blom, H.A.P. (1987). Continuous-discrete filtering for systems with markovian switching coefficients and simultaneous jumps, Proc. 21st Asilomar Conf. Signals Syst. Comp., Pacific Grove, pp.244-248.

Blom, H.A.P., and Bar Shalom, Y. (1988). The interacting multiple model algorithm for systems with markovian switching coefficients, IEEE Trans. Aut. Control, AC-33: 780.

Caines, P.E. and Chen, H.F. (1985). Optimal adaptive LQG control for systems with finite state process parameters,IEEE Trans. Aut. Control, AC-30: 185.

Chang, K.C., and Bar Shalom, Y. (1987). Distributed adaptive estimation with probabilistic data association, Proc. 10th IFAC World Congress, Munich, pp.216-221.

Chizeck, H.J. (1982). Fault-tolerant optimal control, Ph. D. Dissertation, Lab. Inf. Decision Systems, Mass. Inst. Technology, report no. 903-23077.

Chizeck, H.J., Willsky, A.S., and Castanon, D.A. (1986). Discrete-time markovian jump linear quadratic optimal control, Int. J. Control, 43: 213.

Florentin, J.J. (1961). Optimal control of continuous-time markov stochastic systems, J. Electronics Control, 10: 473.

Haaland, K.S., Hutchins, R.G., and Sworder, D.D. (1987). Imaging systems: from corporal to autonomous, Proc. 21st Asilomar Conf. Signals Syst. Comp., Pacific Grove, pp.244-248.

Hijab, O. (1987). Stabilization of Control Systems, Springer-Verlag, New York.

Kazangey, T., and Sworder, D.D. (1971). Effective federal policies for regulating residential housing, Proc. Summer Comp. Simulation Conf., Los Angeles, pp.1120-1128.

Krasovskii, N.N., and Lidskii, E.A. (1961). Analytical design of controllers in systems with random attributes, Parts I-III, Aut. Remote Control, 22: 1021, 1141, 1289.

Ladde, G.S., and Siljak, D.D. (1983). Multiplex control systems: stochastic stability and dynamic reliability, Int. J. Control, 38: 515.

Malhame, R., and Chong, C.Y. (1985). Electric load model synthesis by diffusion approximation in a high order hybrid state stochastic system, IEEE Trans. Aut. Control, AC-30: 854.

Mariton, M., and Bertrand, P. (1986). Improved multiplex control: dynamic reliability and stochastic optimality, Int. J. Control, 44: 219.

Millnert, M. (1982). Identification and control of systems subject to abrupt changes, Ph.D. Dissertation, Linköping Studies in Science and Technology, no.82, Linköping.

Montgomery, R.C. (1983). Reliability considerations in the placement of control system components, Proc. AIAA Guidance and Control Conf., Gatlinburg.

Moore, E.F., and Shannon, C.E. (1956). Reliable circuits using less reliable relays, J. Franklin Inst., 191.

Moose, R.L. (1975). An adaptive state estimation solution to the maneuvering target problem, IEEE Trans. Aut. Control, AC-20: 359.

Samuelson, P.A. (1939). Interactions between the multiplier analysis and the principle of acceleration, Review of Economic Statistics, 21: 75.

Siljak, D.D. (1978). Dynamic reliability using multiple control systems, Proc. 2nd Lawrence Symp. Systems Decision Sciences, Berkeley.

Siljak, D.D. (1980). Reliable control using multiple control systems, Int. J. Control, 31: 303.

Siljak, D.D. (1981). Dynamic reliability of multiplex control systems, Proc. 8th IFAC World Congress, Kyoto, pp.110-115.

Sworder, D.D. (1967). On the control of stochastic systems, Parts I-II, Int. J. Control, 6: 179, 10: 271.

Sworder, D.D. (1969). Feedback control of a class of linear systems with jump parameters, IEEE Trans. Aut. Control, AC-14: 9.

Sworder, D.D. (1976). Control of systems subject to abrupt changes in character, Proc. IEEE, 64: 1219.

Sworder, D.D. (1982). Regulation of stochastic systems with wide-band transfer functions, IEEE Trans. Systems Man Cyber., SMC-12: 307.

Sworder, D.D., and Rogers, R.O. (1983). An LQ solution to a control problem associated with a solar thermal central receiver, IEEE Trans. Aut. Control, AC-28: 971.

Sworder, D.D. (1984). Control of systems subject to small measurement disturbances, Trans. ASME, J. Dyn. Systems Meas. Control, 106: 182.

Sworder, D.D., and Chou, S.D. (1985). A survey of design methods for random parameter systems, Proc. 24th IEEE Conf. Decision Control, Fort Lauderdale, pp.894-899.

Sworder, D.D. (1987). Improved target prediction using an IR imager, Proc. SPIE Conf. Optoelectronics and Laser Appl. in Sciences and Engineering, Los Angeles.

Von Neuman, J. (1956). Automata Studies, in Annals of Mathematics (C.E. Shannon and E.F. Moore, eds), vol.34, Princeton University Press.

Willsky, A.S., and Levy, B.C. (1979). Stochastic stability research for complex power systems, Lab. Inf. Decision Systems, Mass. Inst. Technology, report no.ET-76-C-01-2295.

Wonham, W.M. (1971). Random differential equations in control theory, in Probabilistic Methods in Applied Mathematics (A.T. Bharucha-Reid, ed.), vol.2, Academic Press, New York, p.131.

2

Controllability and Stability

2.1 Introduction

The concepts of controllability and observability introduced by Kalman (1963), have played a key role in the development of system theory. In a sense controllability and observability are the foundations on which design is based : they tell us whether or not elementary guidance, stabilization or estimation objectives would be achievable under ideal conditions. Existence questions are not solely a subject of academic interest and they should be asked at an early stage of a study when plant modifications like adding or displacing sensors and actuators are still possible (Kalman, Falb and Arbib, 1969).

In this chapter we introduce several candidates to play the role of controllability / observability and stabilizability / detectability properties for hybrid models. To simplify the problem we restrict ourselves to the easiest case of linear markov hybrid dynamics, introduced in chapter 1, with

$$\begin{aligned} \dot{x}_t &= A(r_t)\, x_t + B(r_t)\, u_t \\ y_t &= C(r_t)\, x_t \end{aligned} \tag{2.1.1}$$

for the euclidean variables and

$$d\phi_t = \Pi'\, \phi_t\, dt + dM_t \tag{2.1.2}$$

for the indicator ϕ_t of the regime.

Placed early in the book, these results also serve to illustrate some of the specific aspects of hybrid models, due both to their stochastic nature and to coupling of discrete and euclidean variables. The reader is thus cautioned against the deceptively simple structure of the piecewise deterministic model (2.1.1), (2.1.2): though straightforward to describe, this model has some distinctive features that deserve detailed attention.

2.2 Relative Controllability

The notion of controllability of (Kalman, 1963) concerns deterministic systems described by differential or difference equations. For stochastic systems it is difficult to establish a similar notion and several notions can be defined depending on the way randomness is taken into account.

We shall first study a relative notion of controllability, where the target is defined both in $\mathbf{R}^n$ and in probability. The definition is as follows: an initial state x_0 is said *ε-controllable with probability* ρ over the time interval $[t_0, t_f]$ if there exists a control law u(t, x) such that

$$\mathcal{P}\ \{\|\, x_{t_f} \|^2 \geq \varepsilon \mid x_{t_0} = x_0\} \leq 1 - \rho$$

Controllability is said *complete* if this condition holds for any x_0. The basic question behind controllability definitions is the existence of a control action that would transfer the initial state of the system to any desired state. For stochastic systems we must give a probabilistic interpretation to the neighbourhood of the target state and ε-controllability is one of the ways of making this interpretation precise.

Using a Liapunov function argument we obtain a sufficient condition for ε-controllability of a linear hybrid system.

Theorem 2.1

An initial state x_0 of the system (2.1.1), (2.1.2) is ε-controllable with probability ρ over the time interval $[t_0, t_f]$ if there exists N matrices P_i, defined over $[t_0, t_f]$, bounded, symetric and positive definite and N real positive numbers α_i such that the following conditions are satisfed

(i) The P_is satisfy coupled Bernoulli equations

$$\dot{P}_i + A_i'P_i + P_i A_i - P_i B_i B_i'P_i + \sum_{j=1}^{N} \pi_{ij} P_j = 0$$

$$i = 1, N \quad (2.2.1)$$

$$P_i(t_f) = I_n / \alpha_i \qquad \text{for } \alpha_i > 0$$

(ii) At $t = t_0$ the system satisfies

$$x_0' P_i(t_0) x_0 \leq (1 - \rho)\, \varepsilon/\alpha_i \quad (2.2.2)$$

Proof :

The proof is based on the construction of a supermartingale using the P_is and we shall need the infinitesimal generator for the state and regime Markov pair.

Even though controllability results are obtained only for linear state dynamics, the computation of the generator is presented in the general case and then specialized. From the definition of the generator given in appendix 1, it is assumed that the function f ($\dot{x}_{pt} = (f(x_{pt}, u_{pt}, r_t, t))$ is continuous in t, x_p and u_p for r = 1 to N and satisfies the usual growth and smoothness hypotheses. Let $g(x_p, \phi, t)$ be a scalar function of x_p, ϕ, and t also satisfying the conditions of appendix 1 (it is recalled that ϕ is the regime indicator introduced in chapter 1). The generator of the pair (x_{pt}, r_t) under the control action u_{pt} is the operator $\mathscr{L}$ such that

$$\lim_{\Delta \to 0} \frac{1}{\Delta} (E\,\{g(x_{pt}, \phi_t, t) \mid x_{pt^-} = x_p, \phi_{t^-} = \phi, t\} - g(x_{pt^-}, \phi_{t^-}, t^-))$$
$$= g_t(x_p, \phi, t) + \mathscr{L}_u\, g(x_p, \phi, t) \quad (2.2.3)$$

where the index on g indicates partial derivation with respect to the corresponding variable. Under the above conditions $\mathscr{L}_u$ is known to exist and can be computed as follows:

$$g(x_{pt}, \phi_t, t) = g(x_p, \phi, t) + f(x_p, u_p, r, t)'\, g_x(x_p, \phi, t)\, \Delta$$
$$+ \sum_{j=1}^{N} [g(x_{pt}, \phi_t, t) - g(x_{pt^-}, \phi_{t^-}, t^-)]\, d\phi_{tj}$$
$$+ g_t(x_p, \phi, t)\, \Delta + o(\Delta)$$

We take the conditional expectation for $d\phi_t = \Pi' dt + dM_t$ and $r_{t^-} = i$

$$g(x_p, i, t) + f(x_p, u_p, i, t)' g_x(x_p, i, t)\,\Delta$$
$$+\sum_{j=1}^{N} (g(x_p, j, t) - g(x_p, i, t))\, \pi_{ij}\, \Delta$$
$$+ g_t(x_p, i, t)\Delta + o(\Delta)$$

where i in g indicates the value when $\phi_{ti} = 1$. The limit in (2.2.3) is then computed

$$g_t(x_p, \phi, t) + \mathscr{L}_u\, g(x_p, \phi_t, t) = g(x_p, i, t) + f(x_p, u_p, i, t)' g_x(x_p, i, t) + \sum_{j=1}^{N} \pi_{ij}\, g(x_p, j, t)$$

so that the generator produces

$$\mathscr{L}_u\, g(x_p, i, t) = f(x_p, u_p, i, t)' g_x(x_p, i, t) + \sum_{j=1}^{N} \pi_{ij}\, g(x_p, j, t)$$

When f is linear in x_p and u_p, this simplifies to

$$\mathscr{L}_u\, g(x, i, t) = (A_i x + B_i u)' g_x(x, i, t) + \sum_{j=1}^{N} \pi_{ij}\, g(x, j, t) \qquad (2.2.4)$$

with the notations of chapter 1 and (x, u) replacing (x_p, u_p) for the linear model. As shown below this is particularly simple to manipulate when g is quadratic in x (or x_p).

Using this result we now turn to the analysis of the conditions of theorem 2.1. For the P_is defined by (2.2.1) we introduce a scalar function over $[t_0, t_f]$

$$V(x, i, t) = x' P_i(t)\, x \qquad \text{when } \phi_{ti} = 1$$

It has first and second order derivatives with respect to x and a first order derivative with respect to t, $t < t_f$. These derivatives are bounded and V qualifies as a candidate Liapunov function. Define a feedback control law

$$u\,(x, i, t) = -1/2\, B_i'\, P_i(t)\, x_t$$

Under this action and along the trajectories of (2.1.1), (2.1.2) we have from (2.2.4)

$$\mathscr{L}_u V(x, i, t) = x' (\dot{P}_i + A_i'P_i + P_i A_i + \sum_{j=1}^{N} \pi_{ij} P_j - P_i B_i B_i'P_i)x \leq 0 \quad (2.2.5)$$

Hence the non-negative function V is an X_t v R_t-supermartingale (where $X_t = \sigma$ - $\{x_s, t_0 \leq s \leq t\}$ and $R_t = \sigma$ - $\{r_s, t_0 \leq s \leq t\}$ are the σ-algebras generated by the state and regime, respectively). By the supermartingale inequality (see appendix 1).

$$\mathscr{P}\{\sup_{t \in [t_0, t_f]} V(x, r, t) \geq \lambda \mid x_{t0}\} \leq V(x_{t0}, r_{t0}, t_0) / \lambda$$

For the terminal condition of (2.2.1) and condition (ii) this implies

$$\mathscr{P}\{(1/\alpha_i) \| x_{t_f} \|^2 \geq \lambda \mid x_{t0} = x_0\} \leq (1 - \rho) (\varepsilon/\lambda\alpha_i)$$

Now choosing λ as $\lambda = \varepsilon/\alpha_i$ we get

$$\mathscr{P}\{\| x_{t_f} \|^2 \geq \varepsilon \mid x_{t0} = x_0\} \leq 1 - \rho$$

which completes the proof. □

Condition (ii) separates, in the sense of a sufficiency test, controllable and uncontrollable initial states for α_i, ε, ρ and t_f fixed. Indeed defining the hitting probability ρ_H as

$$\rho_H = \mathscr{P}\{\| x_{t_f} \|^2 \leq \varepsilon \mid x_{t0} = x_0\}$$

we deduce from (2.2.2)

$$\rho_H \leq 1 - (\alpha_i/\varepsilon)\, x_0'P_i(t_0)\, x_0 \qquad \text{when } x_{t0} = x_0$$

The bound in the definition of ε-controllability is more and more significant as ρ_H approaches 1. Hence we want to have $(\alpha_i/\varepsilon)\, x_0'P_i(t_0)\, x_0$ small, which can be obtained by selecting $\alpha_i << \varepsilon$. Complete controllability is obtained when (2.2.2) is satisfied for any x_0 but again this is always possible for small α_i. Consequently for linear markov hybrid system ε-controllability and complete ε-controllability are equivalent. We are then not surprised by the proportional form of the control law used in the proof : a smaller α_i increases the gain through the terminal condition $P_i(t_f) = I_n/\alpha_i$ and α_i reflects the familiar trade-off between control accuracy and control expenditure. To hit a smaller target with a higher probability we need to expend more control "energy".

Deterministic controllability of the pairs $[A_i, B_i]$, i = 1 to N, implies the existence of N bounded, symetric and positive definite, Riccati equations,

$$\dot{P}_i = - A_i'P_i - A_i P_i + P_i B_i B_i'P_i - Q_i$$
$$P_i(t_f) = 0 \qquad\qquad i = 1, N$$

for any positive definite matrices Q_i, i = 1,N. Using this set of Riccati equations instead of the Bernoulli equations the proof of the theorem remains valid and (2.2.5) becomes

$$\mathcal{L}_u V(x, i, t) = x' (\sum_{j=1}^{N} \pi_{ij} P_j - Q_i) x$$

We can choose

$$Q_i = \sum_{\substack{j=1 \\ j\neq i}}^{N} \pi_{ij} P_j$$

and we get

$$\mathcal{L}_u V = \pi_{ii} x' P_i x$$

which is negative (recall that $\pi_{ii} = - \sum_{\substack{j=1 \\ j\neq i}}^{N} \pi_{ij} \leq 0$) so that the supermartingale property is preserved.

Hence we see that *deterministic controllability of all the regimes implies ε-controllability of the stochastic hybrid system.* The converse is false.

To illustrate the notion of relative controllability we compute conditions (i) and (ii) of the theorem for a simple model.

Example :

Consider a scalar system $\dot{x} = a(r_t)x + b(r_t)u$ with two regimes and

$$\Pi = \begin{pmatrix} -\pi_1 & \pi_1 \\ \pi_2 & -\pi_2 \end{pmatrix}$$

Using lower case letters the Bernoulli set (2.2.1) is written

$$\dot{p}_1 + 2a_1p_1 - p_1^2 b_1^2 - \pi_1p_1 + \pi_1p_2 = 0 \qquad ; p_1(t_f) = 1/\alpha_1$$

$$\dot{p}_2 + 2a_2p_2 - p_2^2 b_2^2 - \pi_2p_2 + \pi_2p_1 = 0 \qquad ; p_2(t_f) = 1/\alpha_2$$

Selecting parameters as

$$a_1 = -1.,\ a_2 = 0.,\ b_1 = -1.,\ b_2 = 1/\sqrt{2}.,\ \pi_1 = 1.,\ \pi_2 = 2.$$
$$\alpha_2 = \alpha_1/2 = \alpha/2$$

we can compute explicitly

$$p_1(t) = e^{t-tf} / (1 + \alpha - e^{t-tf}) \qquad ; \quad p_2(t) = 2p_1(t)$$

For a uniform distribution of the initial regime ($\mathcal{P}\{r_{t_0} = 1\} = \mathcal{P}\{r_{t_0} = 2\} = 0.5$) the value at $t = t_0$ of the averaged Liapunov function is

$$V_0 = x_0^2 (p_1(0) + p_2(0)) = 1.5x_0^2 e^{t-tf} / (1 + \alpha - e^{t-tf})$$

and the minimal value of $1 - \rho$ is

$$(1 - \rho)_{min} = 3\alpha x_0^2 e^{-tf} / (2\varepsilon(1 + \alpha - e^{-tf}))$$

It corresponds to ε-controllability with the highest probability for fixed α, t_f and ε, as plotted on figure 2.1 for a fixed x_0.

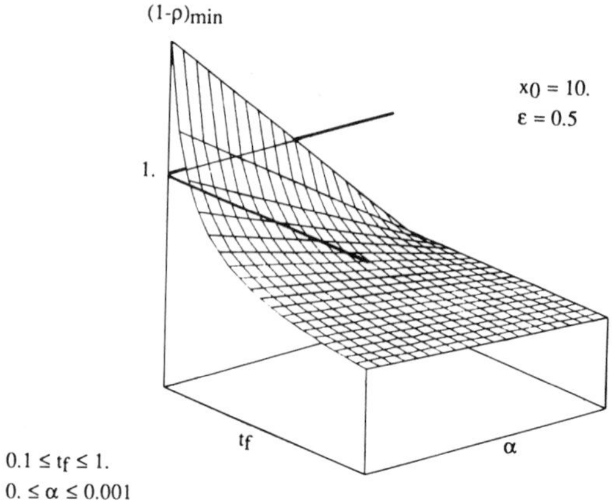

Figure 2.1 - Influence of the final time(t_f) and free parameter (α) on $(1 - \rho)_{min}$.

The larger t_f the easier it is to hit the target and the bound (2.2.2) becomes very tight as ρ tends to 1. A smaller α increases the probability level but at the expense of higher gains which might become unacceptable because of actuator limitations. We can also define the controllability domain : in this scalar example it is simply given by the largest controllable x_0

$$x_{0max} = [(1/3\alpha)\ (2\varepsilon\ e^{-t_f}(1 - \rho)\ (\ (1 + \alpha - e^{-t_f})]^{1/2}$$

Figure 2.2 represents x_{0max} as a function of t_f and ε for three values of ρ.

Larger t_f and ε increase the controllability domain and when it is desired to hit with a very high probability (ρ close to 1) the controllable domain becomes restricted to a small neighbourhood of the origin.

The name *relative* controllability was chosen to stress the difference with the dichotomic yes/no controllability classification in a deterministic setting. As illustrated by the above example, ε-controllability measures a relative and continuous degree of controllability, and can be used to rank candidate actuators configuration. This is especially interesting for flexible structures where there are many degrees of freedom in the choice of components location. Modelling failures as regime transitions, theorem 2.1

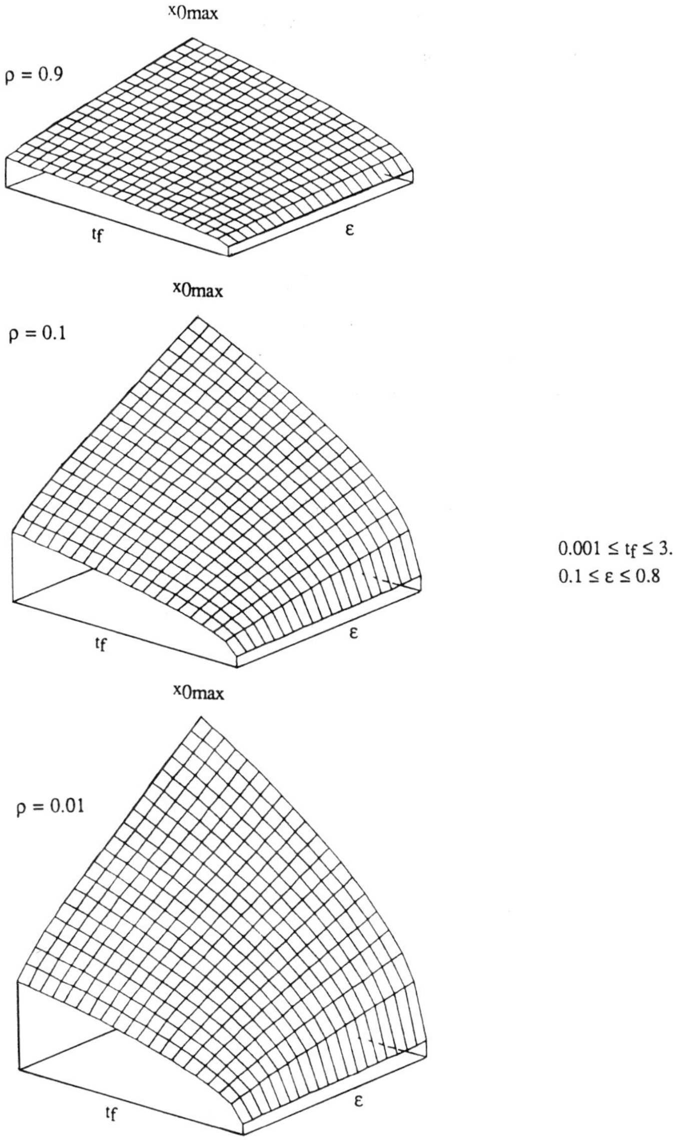

Figure 2.2 - Influence of the final time (t_f) and hitting accuracy (ε) on the controllable domain (x_{0max}).

provides a degree of controllability for the choice of actuators that take into account fault tolerance. This application was explored in (Mariton, 1987b).

2.3 Global Controllability

The above example shows that relative controllability is measure of our capacity to stabilize the system around its equilibrium point. It gives us a construction for a control that drives the initial state into a neighbourhood of the origin at a given time with a given probability. We would like to have a definition that comes closer to the familiar deterministic notion of controllability and better reflects the structure of the dynamic model realization.

As a step in this direction we introduce here a global notion of controllability that has evolved from recurrency concepts of the stochastic processes literature. We will see, however, that this definition has distinctively stochastic aspects and we caution the reader against a too direct transposition of linear deterministic insight.

First we restrict attention to a class of admissible control laws : controllability is discussed with respect to control chosen in the class $\mathcal{U}$ of $X_t \vee R_t$ measurable piecewise continuous functions.

Denoting by $x_t(x_0, u)$ the value at t of the state trajectory under the control u with initial condition $x = x_0$ at time $t = t_0$, we say that the system is *weakly controllable* if $\forall x_0 \in \mathbf{R}^n$, $\forall x_1 \in \mathbf{R}^n$ and $\forall \varepsilon > 0$ there exists an almost surely finite random time T and an admissible control law $u \in \mathcal{U}$ over $[t_0, T]$ such that

$$\mathcal{P}\{\|x_T(x_0, u) - x_1\| \leq \varepsilon\} > 0$$

We say that the system is *controllable* if this probability can be made equal to one and *strongly controllable* if it is weakly controllable and the hitting time

$$T_H = \text{Inf}\ \{T > 0, \|x_T(x_0, u) - x_1\| \leq \varepsilon\}$$

has finite expectation ($E\{T_H\} < \infty$).

The dependence of the state trajectory on the realization of the underlying discrete process is not explicited in the above notations but it is of course the randomness of regime jumps that forces us to define target hitting in probabilistic terms. Notice that controllability, strong and weak controllability are analogues of recurrence, positive recurrence and non degeneracy (or weak recurrence) respectively. Recurency concepts can be illustrated for the regime Markov chain. A Markov chain on S is said *recurrent* if $\forall i \in S$ and $\forall t$, there exists an almost surely finite random time $t' > t$ such that $\mathcal{P}\{r_{t'} = i \mid r_t = i\} = 1$ (it is *weakly recurrent* when this probability is only strictly positive). The chain is said *positive recurrent* if in addition the return time $T_R = \text{Inf}\ \{t > 0, r_t = i, r_0 = i\}$ has finite expectation.

Leaving briefly the hybrid situation, we illustrate the notion of controllability on a simple example. Consider a linear controlled system driven by brownian motion

$$dx_t = Ax_t dt + Bu_t dt + Edw_t$$

It can be shown that this system satisfies the above definition of weak controllability if and only if

$$\text{rank}\ (B, AB, \ldots, A^{n-1}B, E, AE, \ldots, A^{n-1}E) = n$$

The point is that the noise input increases the controllability subspace. The result relies on the non degeneracy of the brownian notion distribution and this reveals a key difference with the deterministic notion of controllability. Further requiring that the part of A not controlled by the "true" control input u_t is stable gives strong controllability. This example serves as a cautionary note to stress that, despite the resemblance of algebraic rank tests like the one above, we should not demand too much to these controllability notions, and certainly not understand the results as naive counterparts to deterministic results.

We shall use the ergodic property for the regime Markov chain. Its significance is that the initial distribution of the regime does not influence the stationary distribution $p_i^\infty > 0$, i = 1 to N (where $p_i^\infty = \lim_{t\to\infty} \mathcal{P}\{r_t = i \mid r_{t_0} = j\}$ for any i, j in S).

A necessary and sufficient condition is obtained in terms of the controllability matrix

$$\mathscr{C} = (\mathscr{C}_1, \mathscr{C}_2, ..., \mathscr{C}_N)$$

where the sub-matrices $\mathscr{C}_i$, i = 1 to N, are themselves assembled from blocks

$$A_{i_1}^{n_1} A_{i_2}^{n_2} \ldots A_{i_N}^{n_N} B_i \quad \text{for} \quad \mathscr{C}_i$$

$$\text{with } i_1, i_2, ..., i_N \in S \quad \text{and} \quad n_1, n_2, ..., n_N \leq n - 1$$

Since we shall be interested only in the range of $\mathscr{C}$ the precise order in which we take the i_k, n_k, k = 1, N, need not be detailed. For example with N = 2 and n = 3 we have

$$\mathscr{C} = (B_1, A_1B_1, A_1^2B_1, A_2B_1, A_2^2B_1, A_2A_1B_1, A_2^2A_1B_1, A_2A_1^2B_1, A_2^2A_1^2B_1,$$
$$B_2, A_2B_2, A_2^2B_2, A_1B_2, A_1^2B_2, A_1A_2B_2, A_1^2A_2B_2, A_1A_2^2B_2, A_1^2A_2^2B_2)$$

The result is as follows.

Theorem 2.2

The linear hybrid model (2.1.1), (2.1.2) with an ergodic Markov chain is strongly controllable if and only if rank $\mathscr{C} = n$.

Proof :

Necessity : with no loss of generality we may assume that the initial state is zero. For a given realization of the regime we indicate by $i_0, i_1, ..., i_k, i_{k+1}$ the regime sequence and by $t_0, t_1, ..., t_k$ the corresponding jump instants ($r_0 = i_0$, $r_{t_0} = i_1, ..., r_{t_k} = i_{k+1}$).
The current state x_t is thus

$$x_t = \int_{t_k}^{t} \exp A_{i_{k+1}} (t - \sigma) B_{i_{k+1}} u_\sigma \, d\sigma + \exp A_{i_{k+1}}(t - t_k) \, x_{t_k}$$

with

$$x_{t_k} = \int_{t_{k-1}}^{t_k} \exp A_{i_k} (t_k - \sigma) B_{i_k} u_\sigma \, d\sigma + \exp A_{i_k}(t_k - t_{k-1}) \, x_{t_{k-1}}$$

Similarly we decompose $x_{t_{k-1}}$ as

$$x_{t_{k-1}} = \int_{t_{k-2}}^{t_{k-1}} \exp A_{i_{k-1}} (t_{k-1} - \sigma) B_{i_{k-1}} u_\sigma \, d\sigma + \exp A_{i_{k-1}}(t_{k-1} - t_{k-2}) \, x_{t_{k-2}}$$

Iterating this decomposition we write x_t as

$$\begin{aligned} x_t = \varphi_{i_0} \int_0^{t_0} \exp A_{i_0} (t_0 - \sigma) B_{i_0} u_\sigma \, d\sigma + \varphi_{i_1} \int_{t_0}^{t_1} \exp A_{i_1}(t_1 - \sigma) B_{i_1} u_\sigma \, d\sigma \\ + \quad \dots \quad + \varphi_{i_k} \int_{t_{k-1}}^{t_k} \exp A_{i_k} (t_k - \sigma) B_{i_k} u_\sigma \, d\sigma \qquad (2.3.1) \\ + \int_{t_k}^{t} \exp A_{i_{k+1}} (t - \sigma) B_{i_{k+1}} u_\sigma \, d\sigma \end{aligned}$$

with realization dependent matrices

$$\varphi_{i_0} = \exp A_{i_{k+1}}(t - t_k) \exp A_{i_k}(t_k - t_{k-1}) \dots \exp A_{i_1}(t_1 - t_0)$$

$$\varphi_{i_k} = \exp A_{i_{k+1}}(t - t_k)$$

But the chain being ergodic we have, for t large enough, the set equalities

$$\{A_{i_0}, A_{i_1}, \dots, A_{i_{k+1}}\} = \{A_1, A_2, \dots, A_N\}$$
$$\{B_{i_0}, B_{i_1}, \dots, B_{i_{k+1}}\} = \{B_1, B_2, \dots, B_N\}$$

and, developing matrix exponentials as in the deterministic case, (2.3.1) shows that x_t belongs to the range of $\mathscr{C}$. The condition rank $\mathscr{C} = n$ is therefore necessary to recover $\mathbf{R}^n$.

Sufficiency : The target x_1 is decomposed N times as $x_1 = y_i + z_i$ with $y_i \in \mathscr{C}_i$ and z_i in its complement. The rank condition guarantees that range $\mathscr{C}_1 + \dots +$ range $\mathscr{C}_N = \mathbf{R}^n$ and the idea of the proof is thus to exhibit a control steering x_0 to a neighbourhood of y_i, the target projected on to range $\mathscr{C}_i$, during a visit to regime i. For any t the contribution to x_t of a visit in regime i_h is

$$\mathscr{C}_{ih} \int_{t_{h-1}}^{t_h} \delta_h(t, t_k, t_{k-1}, \dots, t_h, \sigma)\, u_\sigma \, d\sigma$$

where δ_h is polynomial in its arguments. For $\varepsilon > 0$, $\Delta > 0$ we can find a control law U_Δ defined on $[0, \Delta]$ such that $u_t = U_\Delta(t, y)$ steers any initial condition in range $\mathscr{C}_i$ into an ε neighbourhood of y_i in time Δ.
Now introduce Markov times τ^p_{ai} and τ^p_{bi}, finite with probability one, as

"the chain enters regime i for the p-th time at $t = \tau^p_{ai}$ and subsequently leaves it at $t = \tau^p_{bi}$".

A control law can be defined by

$$u = \begin{cases} U_\Delta(t - \tau^p_{ai}, y(\tau^p_{ai})) & \text{for } \tau^p_{ai} \le t < \tau^p_{bi} \\ 0 & \text{otherwise} \end{cases}$$

where $y(\tau^p_{ai})$ is the projection of $x(\tau^p_{ai})$ onto $\mathscr{C}_i$. But the sojourn times are all exponentially distributed so that the event $\{\tau^p_{ai} + \Delta \le \tau^p_{bi}\}$ has a positive probability, for some p with finite expectation. Repeating this regime by regime proves sufficiency. □

A trivial consequence is that a sufficient condition for weak controllability is the deterministic controllability of a regime (rank $(B_i, A_iB_i, \dots, A_i^{n-1}B_i) = n$ for some $i \in S$). However the hybrid model can be controllable in the above weak sense even when all the regimes are uncontrollable in a deterministic sense, as shown by the following example.

<u>Example</u>

Consider a three regime system in $\mathbf{R}^3$ with $\Pi = \begin{pmatrix} -\pi_1 & \pi_{12} & \pi_{13} \\ \pi_{21} & -\pi_2 & \pi_{23} \\ \pi_{31} & \pi_{32} & -\pi_3 \end{pmatrix}$, $u \in \mathbf{R}$,

and matrices

$$B_1 = (0\ \ 0\ \ 1)' \qquad B_2 = B_3 = (0\ \ 1\ \ 0)'$$

$$A_1 = \begin{pmatrix} 0\,1\,0 \\ 0\,0\,0 \\ 0\,0\,0 \end{pmatrix} \qquad A_2 = \begin{pmatrix} 0\,0\,1 \\ 0\,0\,0 \\ 0\,0\,0 \end{pmatrix} \qquad A_3 = \begin{pmatrix} 0\,0\,0 \\ 0\,0\,0 \\ 0\,0\,1 \end{pmatrix}$$

The ergodicity hypothesis is satisfied and from theorem 2.2 we test

$$\mathscr{C} = (B_1, A_1B_1, A_1^2 B_2, A_2A_1B_1, A_3A_1B_1, B_2 \ldots A_1A_2B_2 \ldots) = 3$$

however

$$\text{rank } (B_1, A_1B_1, A_1^2 B_1) = \text{rank } (B_2, A_2B_2, A_2^2 B_2) = \text{rank } (B_3, A_3B_3, A_3^2 B_3) = 2$$

The controllability subspace defined as range $\mathscr{C}$ is thus larger than the controllability subspaces range $(B_i, A_iB_i, \ldots, A_i^{n-1} B_i)$, i = 1, N. This is because theorem 2.2 allows products of the form $A_j^{h_j} A_i^{h_i} B_i$. However, if we want to reach the target in finite time with probability one, this notion of controllability is not strong enough and we have to turn to the notion of absolute controllability introduced by (Ji and Chizeck, 1987). The stronger condition is then that rank $(B_i, A_iB_i, \ldots, A_i^{n-1} B_i) = n$, i = 1 to N, i.e. deterministic controllability for each regime considered separately. In practice this requirement is reasonable and the plant state models for i = 1 to N are often chosen to satisfy it. Since range $\mathscr{C} \supset$ range $(B_i, A_iB_i, \ldots, A_i^{n-1} B_i)$ we then have a fortiori controllability of the stochastic hybrid model in the sense of theorem 2.2.

2.4 Stochastic Stability

The requirement of stability is a main objective of feedback loop design. This is well known in a deterministic setting but for stochastic systems an additional difficulty is that there are several ways of defining stability (Kozin, 1969). Using (Kushner, 1967) as reference these notions are presented in appendix 2. From the point of view of applications we are interested primarily in sample path and moments stability and we shall study both in the sequel. Our main finding is the role of average dynamics that mix state and regime evolutions. We use fundamental tools from Liapunov theory, namely Liapunov functions and Liapunov exponents. Both reflect the idea that qualitative properties, like stability, of solutions of differential equations (deterministic or stochastic) ought to be determined without actually integrating the differential function.

2.4.1 Liapunov functions

Liapunov functions have been used widely in deterministic as well as stochastic control theory. Roughly speaking the idea is to exhibit a positive scalar function of the state trajectories that decreases as time goes and to infer from that the stability of the system. The difficult step is usually the construction of the Liapunov function itself since, except for linear time invariant systems, no general procedures are known. Rather than fine tuning the function to the structure of the system at hand, the most widely used approach is to guess a simple Liapunov function, almost always a quadratic function of the state, and then to impose constraints on the system so that a negative time derivative can be guaranteed. Obviously this is a rather brutal technique and, not surprisingly, the resulting stability conditions are sometimes very conservative. However it has the merit to be systematic and it has had some success in analysing stability of non-linear or interconnected systems (see e.g. Siljak, 1978).

For stochastic systems, the idea of negative time derivative has to be refined and it is the notion of supermartingale that is called for as shown by the following theorem, adapted from (Kats and Krasovskii, 1960).

Theorem 2.3

A hybrid system (x_t, r_t) with infinitesimal generator $\mathscr{L}$ is exponentially stable in mean square (ESMS) if there exists a Liapunov function V(x, r, t) such that

(i) $V(0, r, t) = 0$

(ii) For a fixed r, V(x, r, t) is continuous and has bounded first derivatives with respect to x and t.

(iii) $c_1 \| x \|^2 \leq V(x, r, t) \leq c_2 \| x \|^2$

(iv) $\mathscr{L} V(x, r, t) \leq - c_3 \| x \|^2$ (2.4.1)

for c_1, c_2 and c_3 positive real numbers. Denoting by $X_t \vee R_t$ the state and regime σ-algebra $\sigma\text{-} \{x_s, r_s, t_0 \leq s \leq t\}$, V is then a non negative $X_t \vee R_t$-supermartingale.

Proof :

See (Kats and Krasovskii, 1960) for the classical proof where the generator in (2.4.1) was replaced by the "generalized derivative". In modern terms the supermartingale property of V, a direct consequence of (2.4.1), expresses the same result in a more compact form, as was first noticed in (Bucy, 1965), see also (Kushner, 1967). □

It is worth noting that theorem 2.3 is limited to moment stability and, more precisely, to mean square stability. The reason for focusing on this aspect of stochastic stability is partially negative in the sense that it is dictated by our ability to exhibit the corresponding Liapunov functions : as it will be seen below, quadratic functions of x are easily constructed to meet the conditions of theorem 2.3 and this simplicity is certainly one motivation for paying special interest to mean square stability. In practice it is nevertheless clear that one cannot be satisfied by an ESMS property. When observing a system in the real world it is a sample path, and not a "mean path", that is observed so that the property of interest should rather be almost sure stability. There are however good reasons to be satisfied for the present time with a mean square study : first the design of optimal regulators for jump linear systems is based on an averaged quadratic cost (see chapter 3) and it will be interesting to relate optimization to the associated, mean square, stabilization. Second the relationship between moments and almost sure stability is such that the ESMS conditions analyzed here will turn out to provide also sufficient conditions for almost sure stability (see § 2.4.2 below).

Using the above theorem it is desired to analyze stabilization of a jump linear system by a linear state feedback control law with regime dependent gain

$$u_t = \Gamma_i \, x_t \qquad \text{when } r_t = i$$

Precisely the system will be said *stochastically stabilizable*, in the ESMS sense, if there exists gains $\Gamma_1, \Gamma_2, \ldots, \Gamma_N$ such that the closed-loop dynamics $\tilde{A}_i = A_i + B_i\Gamma_i$, $i = 1,N$, with jump rates given by Π is ESMS.

A sufficient condition is

<u>Theorem 2.4</u>

A jump linear system $[A_i, B_i, i = 1,N, \Pi]$ is stochastically stabilizable if there exists feedback gains Γ_i, $i = 1,N$, such that for positive definite matrices Q_i, $i = 1,N$, the solutions P_i, $i = 1,N$, of the set of coupled Liapunov equations

$$\tilde{A}_i' P_i + P_i \tilde{A}_i + Q_i + \sum_{j=1}^{N} \pi_{ij} P_j = 0 \tag{2.4.2}$$

are positive definite.

Proof :
For the conditions of the theorem the quadratic function

$$V(x, i, t) = x' P_i x$$

obviously satisfies (i), (ii) and (iii) of theorem 2.3 to be considered as a candidate Liapunov function. Under state feedback the state dynamics are $\dot{x}_t = \tilde{A}(r_t)\, x_t$ with $\tilde{A}(r_t) = A(r_t) + B(r_t)\, \Gamma(r_t)$ and, as previously, we have

$$\mathscr{L}V = x_t' \left(\tilde{A}_i' P_i + P_i \tilde{A}_i + \sum_{j=1}^{N} \pi_{ij} P_j\right) x_t \qquad \text{when } r_t = i$$

so that for the P_is solutions of (2.4.2)

$$\mathscr{L}V = -\, x_t' Q_i\, x_t \qquad \text{when } r_t = i$$

which ensures condition (iv) of theorem 2.3 for positive definite Q_is. □

Compared to the usual Liapunov equation characterizing stable linear deterministic systems, the occurrence of jumps introduces coupling ($\sum_j \pi_{ij} P_j$) and (2.4.2) is indeed associated to both the continuous (A_i, B_i, $i = 1,N$) and discrete (Π) parts of the hybrid model.

2.4.2 Liapunov exponents

Liapunov's name remains most strongly attached to the energy like functions that have been a favorite tool of researchers to assess stability of systems, both deterministic and stochastic. This path was explored in the previous section for our hybrid models. However Liapunov introduced, for deterministic systems, another method, namely to compare the solution of the differential equation to exponential functions by means of characteristic exponents, the so-called Liapunov exponents. Consider the deterministic equation

$$\dot{x}_t = f(x_t)$$

in some neighbourhood of $x_{t_0} = x_0$. For a sufficiently smooth f it might be possible to define

$$\lambda(x_0) = \lim_{t \to \infty} \frac{1}{t} \log \| x_t \|$$

Clearly λ gives the asymptotic growth rate and for f linear, $f(x_t) = Ax_t$, λ is readily computed as $\lambda = \sigma(A)$, the real part of the dominant eigenvalue of A excited by x_0. The sign of λ then characterizes the stability of f around x_0 with locally stable directions when $\lambda(x_0) < 0$ and unstable ones when $\lambda(x_0) > 0$.

It is only recently that this approach has been extended to stochastic systems (Arnold, 1984). Denoting by ω the hazard, the solution of a stochastic differential equation originating at x_0 can be analyzed by means of the sample path Liapunov exponent

$$\lambda(\omega, x_0) = \lim_{t \to \infty} \frac{1}{t} \log \| x_t(x_0, \omega) \| \qquad (2.4.3)$$

Obviously almost sure stability is intimately related to the sign of $\lambda(\omega, x_0)$ and indeed it can be shown (Arnold et al., 1986) that a necessary and sufficient condition for almost sure stability is

$$\sup_{x_0 \neq 0} \lambda(\omega, x_0) < 0 \quad \text{a.s.}$$

Characteristic exponents can also be defined for the moments of $x_t(x_0, \omega)$

$$\lambda_p(x_0) = \lim_{t \to \infty} \frac{1}{t} \log E\{\| x_t(x_0, \omega)\|^p\} \qquad (2.4.4)$$

with again a pth moment stability condition depending on the sign of $\lambda_p = \sup_{x_0 \neq 0} \lambda_p(x_0)$.

The purpose of this section is to compute (2.4.3) and (2.4.4) for jump linear systems and to use the results to study the interplay between moments and almost sure stability.

Moments propagation is considered first. The pth moments ($p \in \mathbf{N}$) are expressed using the p-fold Kronecker product of x_t with itself $x_t^{(p)} = x_t \otimes x_t \otimes \ldots \otimes x_t$, p times, and

$$X_{ti}^{p} = E\{x_t^{(p)} \phi_{ti}\} \qquad , i = 1,N$$

where ϕ_t is the regime indicator. In terms of the X_{ti}^{p}, i = 1 to N, one has

$$E\{x_t^{(p)}\} = \sum_{i=1}^{N} X_{ti}^{p}$$

It is therefore desired to propagate the X_{ti}^{p}, i = 1, N with time. For p = 1 this is just

$$\dot{x}_t^{(1)} = \dot{x}_t = A(r_t)\, x_t = A(r_t)\, x_t^{(1)} \tag{2.4.5}$$

and, for the Markov chain model,

$$X_{t+\Delta i}^{1} = \Delta\, A_i\, X_{ti}^{1} + \sum_{j=1}^{N} \mathscr{P}\{r_{t+\Delta} = j \mid r_t = i\}\, X_{tj}^{1} + o(\Delta)$$

so that

$$\dot{X}_{ti}^{1} = A_i\, X_{ti}^{1} + \sum_{j=1}^{N} \pi_{ij}\, X_{tj}^{1} \qquad , i = 1,N \tag{2.4.6}$$

The set of N linear equations (2.4.6) can be written as a single vector equation with $X_t^1 = [X_{t1}^{1\prime}, X_{t2}^{1\prime}, \ldots, X_{tN}^{1\prime}]'$, $X_t^1 \in \mathbf{R}^{Nn}$

$$\dot{X}_t^{1} = F_1 X_t^{1} \qquad \text{for } F_1 = \text{diag}\,(A_i)_{\,i=1,N} + \Pi' \otimes I_n \tag{2.4.7}$$

For higher order moments, the equation satisfied by $x_t^{(p)}$ is first derived and then the result obtained for the first moment can be applied. For example when p = 2

$$\dot{x}_t^{(2)} = [A(r_t) \otimes I_n + I_n \otimes A(r_t)]\, x_t^{(2)}$$

which gives, as in (2.4.7) with $X_t^2 \in \mathbf{R}^{Nn^2}$

$$\dot{X}_t^2 = F_2 X_t^2 \qquad \text{for } F_2 = \text{diag}\,(A_i \otimes I_n + I_n \otimes A_i)_{i=1,N} + \Pi' \otimes I_{n^2}$$

The general expression is

$$\dot{X}_t^p = F_p\, X_t^p$$

for $X_t^p \in \mathbf{R}^{Nn^p}$ and

$$F_p = \text{diag}\,\Big(\sum_{k=1}^{p} I_n{}^{(k-1)} \otimes A_i \otimes I_n{}^{(p-k)}\Big)_{i=1,N} + \Pi' \otimes I_{n^p} \tag{2.4.8}$$

with $X^p_{t=t_0} = [X^{p\,\prime}_{01}, X^{p\,\prime}_{02}, \dots, X^{p\,\prime}_{0N}]'$, $X^p_{0i} = x_0{}^{(p)} \mathcal{I}\{r_{t_0} = i\}$, $i = 1,N$. From the above propagation equation it is straightforward to derive a stability result by explicitly computing the pth Liapunov exponent.

<u>Theorem 2.5</u>

The pth Liapunov exponent of a linear system with markovian jumps is given by $\lambda_p = \sigma(F_p)$ where $\sigma(F_p)$ is the real part of the dominant eigenvalue of the averaged dynamics matrix (2.4.8), and the pth moment is stable iff $\sigma(F_p) < 0$.

<u>Proof</u>:

From the above computation it follows that the pth moment propagates with time as a linear combination of exponentials in the eigenvalues of F_p, hence $\lambda_p \le \sigma(F_p)$. But it cannot happen that an eigenvalue of F_p does not contribute : $E\,\{x_t^{(p)}\} = He^{F_p t} X^p_{t_0}$ with $H = [I_{n^p}, I_{n^p}, \dots, I_{n^p}]$ and the generic observability of [F, H] and the sup with respect to x_0 in the stability test ensures $\lambda_p = \sigma(F_p)$. □

For the controlled dynamics

$$\dot{x}_t = A(r_t)\, x_t + B(r_t)\, u_t$$

under linear regime dependent feedback laws, $u_t = \Gamma_i\, x_t$ when $r_t = i$, this theorem also provides a condition for feedback stabilizability as $\sigma(\tilde{F}_p) < 0$ where $\tilde{F}_p$ is defined as in (2.4.8) with A_i replaced by $\tilde{A}_i = A_i + B_i\Gamma_i$, i = 1 to N. In fact selecting the gains Γ_i, i = 1 to N, to modify the spectrum of F_p into that of $\tilde{F}_p$ is formally similar to a decentralized control problem where one tries to stabilize a system by acting on the subsystems (here the diagonal blocks of F_p) for fixed interconnections (here the off-diagonal blocks of F_p).

Almost sure stability is analyzed next, using the idea of Coppel's dichotomy (Blankenship,1977). The trajectory x_t can be bracketed as

$$\| x_0 \| \exp - \int_{t_0}^{t} \sigma\,(-A(r_s))\, ds \leq \| x_t \| \leq \| x_0 \| \exp \int_{t_0}^{t} \sigma\,(A(r_s))\, ds$$

with σ(-) as before the real part of the dominant eigenvalue. Sufficient conditions for almost sure stability and instability are then

Theorem 2.6

The system with an ergodic Markov chain is almost surely stable (resp. unstable) if $\sum_{i=1}^{N} p_i^\infty\, \sigma(A_i) < 0$ (resp. $\sum_{i=1}^{N} p_i^\infty\, \sigma(-A_i) < 0$) where p_i^∞ is the asymptotic probability of mode i.

Proof:

The sample path condition on $\lambda = \sup_{x_0 \neq 0} \lambda(\omega, x_0)$ can be enforced through the upper and lower bounds

$$\lim_{t\to\infty} (1/t) \left(- \int_{t_0}^{t} \sigma\,(-A(r_s))\, ds\right) \leq \lambda \leq \lim_{t\to\infty} (1/t) \left(\int_{t_0}^{t} \sigma\,(-A(r_s))\, ds\right)$$

But the Markov chain is assumed ergodic and the time averages are computed as

$$\lim_{t\to\infty} (1/t) \left(- \int_{t_0}^{t} \sigma (A(r_s))\, ds \right) = \sum_{i=1}^{N} p_i^{\infty} \sigma (A_i)$$

and

$$\lim_{t\to\infty} (1/t) \left(- \int_{t_0}^{t} \sigma (-A(r_s))\, ds \right) = \sum_{i=1}^{N} p_i^{\infty} \sigma (-A_i)$$

so that $- \sum_{i=1}^{N} p_i^{\infty} \sigma (-A_i) \leq \lambda \leq \sum_{i=1}^{N} p_i^{\infty} \sigma (A_i)$ and the theorem follows from the general result of (Arnold and Wihstutz, 1986). □

As opposed to theorem 2.5, theorem 2.6 provides only sufficient conditions and there remains uncertainty regarding a.s. stability when both conditions are violated as with the following example :

For $N = 2$, $n = 2$,
$A_1 = \text{diag}\,(1, -2)$, $A_2 = I_2$, $\pi_{12} = 2$ $\pi_{21} = 1$

one has

$$p_1^{\infty} \sigma (A_1) + p_2^{\infty} \sigma (A_2) = (2/3) \times 1 + (1/3) \times 1 = 1 > 0$$

and

$$p_1^{\infty} \sigma (-A_1) + p_2^{\infty} \sigma (-A_2) = (2/3) \times 2 - (1/3) \times 1 = 1 > 0$$

The main advantage of the above conditions are therefore their simplicity and transparent interpretation in terms of averaged dynamics but tighter, and more complex, conditions might be needed in some cases. This could be accomplished with the technique of (Loparo and Blankenship, 1986).

2.4.3 Discussion

Liapunov functions and exponents have led to different stability conditions, but a common feature of these conditions is that they display an interaction between the continuous and discrete parts of the hybrid dynamics. To better understand this interaction, and the relationship between moments and sample path stability, several examples are now discussed.

First the condition of theorem 2.4 is illustrated by three examples.

Example 1: Deterministic stabilizability of the regimes is not necessary for stochastic stabilizability.

Consider a system in $\mathbf{R}^2$ with two regimes (n = N = 2) and

$$A_1=\begin{bmatrix} 1 & -1 \\ 0 & -2 \end{bmatrix} \quad ; \quad B_1=\begin{bmatrix} 0 \\ 0 \end{bmatrix}$$

$$A_2=\begin{bmatrix} -2 & 1 \\ 0 & 1 \end{bmatrix} \quad ; \quad B_2=\begin{bmatrix} 0 \\ 0 \end{bmatrix}$$

and a symetric transition matrix $\Pi = \begin{bmatrix} -\pi & \pi \\ \pi & -\pi \end{bmatrix}$. Matrices B_1 and B_2 being zero the unstable eigenvalue at 1 of A_1 and A_2 cannot be compensated so that neither regime 1 nor regime 2 are stabilizable in a deterministic sense. However, the condition of theorem 2.4, for asymptotically fast jumps ($\pi \rightarrow \infty$), is satisfied if $\tilde{A}_1 + \tilde{A}_2$ has stable eigenvalues. But here $\tilde{A}_1 + \tilde{A}_2 = -I$ it follows that $[A_1, A_2, \Pi]$ is stochastically stabilizable in the ESMS sense. This example demonstrates that regime deterministic stability is not necessary to have stochastic stability of the hybrid system. There appears some "averaged dynamics" as with the Liapunov exponents approach : noting that $p_1^\infty = p_2^\infty = 0.5$ for this example, $\tilde{A}_1 + \tilde{A}_2$ is just two times $p_1^\infty \tilde{A}_1 + p_2^\infty \tilde{A}_2$.

Example 2: Deterministic stabilizability of the regimes is not sufficient for stochastic stabilizability.

Consider again n = N = 2 and

$$A_1=\begin{bmatrix} -1 & 10 \\ 0 & -1 \end{bmatrix} \quad ; \quad B_1=\begin{bmatrix} 0 \\ 0 \end{bmatrix}$$

$$A_2=\begin{bmatrix} -1 & 0 \\ 10 & -1 \end{bmatrix} \quad ; \quad B_2=\begin{bmatrix} 0 \\ 0 \end{bmatrix}$$

with Π as in example 1. Both A_1 and A_2 are stable (eigenvalues -1,-1) but for $\pi \rightarrow \infty$ the condition of theorem 2.4 leads as before to the eigenvalues of $\tilde{A}_1 + \tilde{A}_2$. Now these are 8 and -12 and the stochastic stability condition is not satisfied (unstable eigenvalue at 8). Here the averaging reveals that regime deterministic is not sufficient to guarantee stochastic stability of the hybrid system.

Example 3:

Consider a scalar system with five regimes (n=1, N=5) described by

$$\dot{x}_t = a_i x_t + u_t \ ; \quad u_t = \gamma_i x_t \qquad \text{when } r_t = i$$

and a matrix of transition rates

$$\Pi = \begin{bmatrix} -\pi_{12} & \pi_{12} & 0 & 0 & 0 \\ 0 & -(\pi_{23}+\pi_{25}) & \pi_{23} & 0 & \pi_{25} \\ 0 & 0 & -\pi_{34} & \pi_{34} & 0 \\ 0 & 0 & \pi_{43} & -\pi_{43} & 0 \\ 0 & 0 & 0 & 0 & 0 \end{bmatrix}$$

Using the regime classification for Markov chains explained in appendix 1, it appears that Π corresponds to two transient regimes (r = 1, 2), a closed communicating class (r=3, 4) and an absorbing regime (r = 5). The structure of this chain is displayed on figure 2.3.

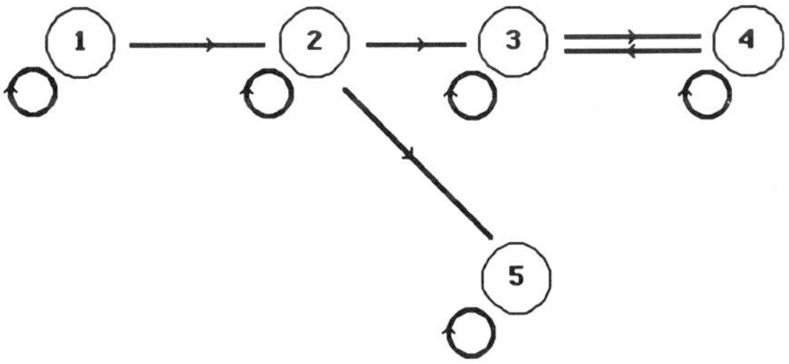

Figure 2.3 - Graph of the chain for example 3.

The condition of theorem 2.4 can then be written for different categories of regimes.

- Absorbing regime (r = 5) :

The condition reduces to

$$a_5 + \gamma_5 < 0$$

in other words to the deterministic stability condition for r = 5. This is expected because 5 being absorbing the stochastic nature of the system degenerates once it enters r = 5 and the stochastic condition and its deterministic counterpart coalesce.

- Closed communicating class (r = 3, 4) :

Letting $\pi_{34} = \pi_{43} = \pi$ and $\pi \to \infty$ the condition is

$$-(2 + \gamma_3^2 + \gamma_4^2) / 2(a_3 + \gamma_3 + a_4 + \gamma_4) > 0$$

and averaging is again apparent. It may happen that $a_3 + \gamma_3 + a_4 + \gamma_4 < 0$, ensuring ESMS trajectories, even if $a_3 + \gamma_3 > 0$, provided $a_4 + \gamma_4$ is "negative enough" ($a_4 + \gamma_4 < -(a_3 + \gamma_3)$). In other words stochastic stability is not jeopardized by the existence of deterministically unstable regimes within a communicating class, provided other regimes of the class compensate.

- Transient regimes (r = 1, 2) :

For regime 2 the condition is

$$a_2 + \gamma_2 - (\pi_{23} + \pi_{25}) / 2 < 0$$

Rates π_{23} and π_{25} being positive this can happen with positive $a_2 + \gamma_2$, that is for a deterministically unstable regime 2. A similar fact holds for r = 1. The interpretation is that the hybrid system can tolerate unstable regimes $a_1 + \gamma_1 > 0$, $a_2 + \gamma_2 > 0$ provided their average life time is short enough (π_{12}, $\pi_{23} + \pi_{25}$ large).

While Liapunov functions prove useful in studying mean square stability, the Liapunov exponents of § 2.4.2 reveal the relationship between moments and sample path stability. This is now studied in details by means of an example.

Example 4 :

Consider a scalar system with two modes (n = 1, N = 2) and transitions rates $\pi_{12} = -\pi_{11} = \pi_1$ and $\pi_{21} = -\pi_{22} = \pi_2$. The average life time of mode i is $1/\pi_{ii}$, i = 1, 2 and $p_1^\infty = \pi_2 / (\pi_1 + \pi_2)$. From theorem 2.5 stability of the pth moment is ruled by the eigenvalues of the matrix

$$\begin{bmatrix} pa_1 - \pi_1 & \pi_2 \\ \pi_1 & pa_2 - \pi_2 \end{bmatrix}$$

For this simple situation, it is possible to explicitly solve the characteristic polynomial of F_p to compute λ_p. Depending on the values of the model parameters (a_1, a_2, π_1 and π_2), only four situations may occur. Almost sure stability is obtained from theorem 2.6 when $p_1^\infty \sigma(A_1) + p_2^\infty \sigma(A_2) = p_1^\infty a_1 + p_2^\infty a_2$ is negative and figure 2.4 plots λ_p as a function of p for this case.

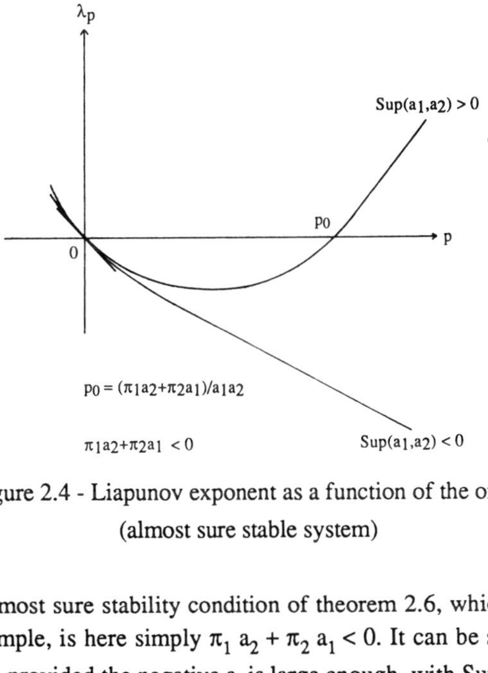

Figure 2.4 - Liapunov exponent as a function of the order.
(almost sure stable system)

The almost sure stability condition of theorem 2.6, which is also necessary for this scalar example, is here simply $\pi_1 a_2 + \pi_2 a_1 < 0$. It can be satisfied either with Sup $(a_1, a_2) < 0$ or, provided the negative a_i is large enough, with Sup $(a_1, a_2) > 0$. When Sup $(a_1, a_2) > 0$, figure 2.4 shows that large order moments ($p > p_0 > 0$ with $p_0 = (\pi_1 a_2 + \pi_2 a_1) / (a_1 a_2)$) are unstable, their Liapunov exponent λ_p being positive. This corresponds to the occurrence of large deviations when, even though x_t decays to zero exponentially with probability one, some rare large excursions occur so that for p large enough $E\{\| x_t \|^p\}$ actually grows. When Sup $(a_1, a_2) < 0$, it is observed on figure 2.4 that all the moments are stable ($\lambda_p < 0, \forall p > 0$). It can be noted that the condition for the stability of all moments (Sup $(a_1, a_2) < 0$) is more demanding than the almost sure stability condition ($\pi_2 a_1 + \pi_1 a_2 < 0$), again the averaging between state and regime dynamics is manifest and the weaker condition can be satisfied with regime 1 unstable provided regime 2 is stable enough ($a_2 < 0$) and the system does not stay, on the average, too long in regime 1 (π_1 large).

The two remaining cases correspond to systems that are almost sure unstable ($p_1^\infty a_1 + p_2^\infty a_2 > 0$) with again the sign of Sup (a_1, a_2) as the frontier. To conclude this example, the derivative of λ_p at $p = 0$ can be formally evaluated as

$$\lambda'_{p=0} = (\pi_1 a_2 + \pi_2 a_1) / (\pi_1 + \pi_2)$$

which is equal to the almost sure Liapunov exponent $\lambda = p_1^\infty a_1 + p_2^\infty a_2$. In other words, the slope at the origin of the moment Liapunov exponent (as a function of the order) is given by the almost sure Liapunov exponent. The existence of a stable positive moment (e.g. $p = 2$, mean square stability which can be established using a Liapunov function condition) implies almost sure stability ($\lambda = \lambda'_{p=0} < 0$, see figure 2.4).

The condition for stability of high order moments is easily extended to more general examples : indeed as p grows, the matrix F_p of theorem 2.5 becomes increasingly diagonal dominant so that its spectrum consists of eigenvalues

$$\mu_{k_1}^{i_1} + \mu_{k_2}^{i_2} + \dots + \mu_{k_L}^{i_L} \qquad \text{for } L = Nn^p$$

where k_1 , k_2, ... $\in$ {1, 2, ... , N}, i_1, i_2, ... $\in$ {1, 2, ... , N} and the μ_k^i, $k = 1, N$, are the eigenvalues of A_i, $i = 1, N$. Therefore the condition Sup $(a_1, a_2) < 0$ obtained for the scalar example is extended to the general case ($N > 2$, $n > 1$) as Sup $(\sigma(A_1), \sigma(A_2), \dots, \sigma(A_N)) < 0$.

2.5 Dual notions : Observability and Detectability

As in the deterministic setting there is a duality between the input and the output of the system or between control and observation. Using the analysis on controllability and stabilizability in previous paragraphs we now propose the corresponding notions of observability and detectability.

To the notion of ε-controllability duality associates ε-observability. A system is ε-controllable if the initial state can be transferred in a stochastic sense made precise in § 2.2 to any desired state within a finite time by some control actions. The definition of ε-observability similarly captures the existence and accuracy of an estimator reconstructing the state from output measurements : a system is said *ε-observable* with probability ρ if there exists a state estimate, noted $\hat{x}_t$, such that

$$\mathcal{P}\{ \| \hat{x}_{t_f} - x_{t_f} \|^2 \geq \varepsilon \mid \hat{x}_{t_0} - x_{t_0} = \hat{x}_0 - x_0 \} \leq 1 - \rho$$

Using a supermartingale argument we obtain sufficient conditions for ε-observability in the manner of theorem 2.1 with the Bernoulli equations (2.2.1) replaced by

$$\dot{P}_i + A_i P_i + P_i A_i' - P_i C_i' C_i P_i + \sum_{j=1}^{N} \pi_{ij} P_j = 0$$

$$i = 1,N$$

$$P_i(t_f) = I_n / \alpha_i$$

The notion of strong controllability of § 2.3 leads to a dual definition of strong observability and, with an ergodic Markov chain for the regime, the observability matrix is formed as

$$\mathcal{O} = (\mathcal{O}_1', \mathcal{O}_2', \dots, \mathcal{O}_N')'$$

where the sub-matrices $\mathcal{O}_i$, i = 1 to N, are themselves assembled from blocks like

$$C_i A_{i_1}^{n_1} A_{i_2}^{n_2} \dots A_{i_N}^{n_N} \qquad \text{for } \mathcal{O}_i$$

with $i_1, i_2, \dots, i_n \in S$ and $n_1, n_2, \dots, n_N \leq n - 1$

A necessary and sufficient condition for stochastic strong observability is rank $\mathcal{O} = n$. The proof mimicks that of theorem 2.2.

Finally the notion of stochastic detectability is dual to that of stochastic stabilizability in § 2.4.1. It characterizes the $[A_i, C_i, i = 1,N, \Pi]$ model as *stochastically detectable* if its dual $[A_i' \; C_i', i = 1,N, \Pi]$ is stochastically stabilizable. Analogous to theorem 2.4 we obtain sufficient conditions

Theorem 2.6

The system $[A_i, C_i, i = 1,N, \Pi]$ is stochastically detectable if there exist observer gains K_i, i = 1, N, such that for positive definite matrices Q_i, i = 1, N, the solutions P_i, i = 1, N of the set of coupled Liapunov equations

$$\tilde{A}_i P_i + P_i \tilde{A}_i' + Q_i + \sum_{j=1}^{N} \pi_{ij} P_j = 0$$

are positive definite.

Proof :

Refer to the proof of theorem 2.4. □

In chapter 3 we shall use stochastic stabilizability and detectability together in solving the asymptotic JLQ regulator.

2.6 Notes and References

This chapter presents various attempts at defining counterparts of the familiar and instrumental controllability / observability, stabilizability / detectability notions of deterministic system theory. Through these studies it becomes clear that stochastic systems, and, in particular, hybrid systems have highly distinctive features that make it risky to extrapolate too far deterministic intuition. Within this chapter this was best illustrated by the notion of averaged dynamics that intimately mix the plant state and plant regime evolution.

Most of the material in this chapter is based on the papers (Mariton and Bertrand, 1985, 1986, Mariton, 1986, 1987a). On relative controllability we used the work of (Sunahara et al., 1974, 1975, Klamka and Socha, 1977, Socha, 1984). On strong controllability, the corresponding recurrency concepts are those of (Arnold and

Klieman, 1982) adapted to the control context as in (Zabcyk, 1981, Ehrhardt and Klieman, 1982). A related analysis is (Ji and Chizeck, 1987) from which we borrow the absolute controllability concept mentioned in § 2.3. An overview of stability in stochastic systems is given in appendix 2, including reference to the comprehensive studies of (Bertram and Sarachik, 1959, Kozin, 1969).

References

Arnold, L. (1984). A formula connecting sample path and moment stability of linear stochastic systems, SIAM J. Appl. Math., 44 : 783.

Arnold, L., and Klieman, W. (1982). Qualitative theory of stochastic systems, in Probabilistic Analysis and Related Topics (A.T. Bharucha-Reid, Ed.), vol. 3, Academic Press, New York, p.1.

Arnold, L. , and Wihstutz, V. , Eds (1986). Liapunov Exponents , Proc. Workshop Bremen University, Lecture Notes in Mathematics, Vol. 1186, Springer Verlag, Berlin.

Bertram, J.E. , and Sarachik, P.E. (1959). Stability of circuits with randomly time varying parameters, Trans. IRE, PGIT-5 : 260.

Blankenship, G.L. (1977). Stability of linear differential equations with random coefficients, IEEE Trans. Aut. Control, AC-22 : 834.

Bucy, R.S. (1965). Stability and positive supermartingales, J. Diff. Eqs, 1 : 151.

Ehrhardt, M., and Klieman, W. (1982). Controllability of linear stochastic systems, Syst. Control Letters, 2 : 145.

Ji, Y. , and Chizeck, H.J. (1987). Controllability, observability and jump linear quadratic problem in continuous time, Proc. 26 th IEEE Conf. Decision Control, Los Angeles, pp. 329-331.

Kalman, R.E. (1963). Mathematical description of linear systems, SIAM J. Contr. Opt., 1: 152.

Kalman, R.E. , Falb, P.L. , and Arbib, M.A. (1969). Topics in Mathematical Systems Theory, Mac Graw Hill, New York.

Kats, I.I. , and Krasovskii, N.N. (1960). On the stability of systems with random attributes, J. Appl. Math. Mech., 24 : 1255.

Klamka, J. , and Socha, L. (1977). Some remarks about stochastic controllability, IEEE Trans. Aut. Control, AC-22 : 880.

Kozin, F. (1969). A survey of stability of stochastic systems, Automatica, 5 : 95.

Kushner, H. (1967). Stochastic Stability and Control, Academic Press, New York.

Loparo, K.A. , and Blankenship, G.L. (1986). Almost sure instability of a class of linear stochastic system with jump process coefficients, in Liapunov Exponents (L. Arnold and V. Wihstutz, Eds), Lecture Notes in Mathematics, Vol. 1186, Springer Verlag, Berlin, p. 160.

Mariton, M. , and Bertrand, P. (1985). Comportement asymptotique de la commande pour les systèmes linéaires à sauts markoviens, C.R. Acad. Sciences, Paris, 301 : 683.

Mariton, M. , and Bertrand, P. (1986). Asymptotic behaviour of jump linear systems in continuous time, Proc. 7th INRIA Conf. Analysis and Optim. Systems, Lions, J.L. and Bensoussan, A., Eds. , Antibes, pp. 483-494.

Mariton, M. (1986). On controllability of linear systems with stochastic jump parameters, IEEE Trans. Aut. Control, AC-31 : 680.

Mariton, M. (1987a). Stochastic controllability of linear systems with markovian jumps, Automatica, 23 : 783.

Mariton, M. (1987b). Placement of failure-prone components on flexible structures : a degree of controllability approach, Proc. American Control Conf., Minneapolis, pp. 1883-1884.

Siljak, D.D. (1978). Large-Scale Dynamic Systems : Stability and Structure, North Holland, New York.

Socha, L. (1984). Some remark about stochastic controllability for linear composite systems, IEEE Trans. Aut. Control, AC-29 : 60.

Sunahara, Y. , Kabeuchi, T. , Yasada, Y. , Aihara, S. , and Kishino, K. (1974). On stochastic controllability of non linear systems, IEEE Trans. Aut. Control, AC-19 : 49.

Sunahara, Y. , Aihara, S., and Kishino, K. (1975). On the stochastic observability and controllability for non linear systems, Int. J. Control, 22 : 65.

Zabcyk, J. (1981). Controllability of stochastic linear systems, Syst. Control Letters, 1 : 25.

3
Control Optimization

3.1 Problem Formulation

A control law can be conceptually split into two parts, the guidance law and the stabilization law. Guiding terms are most often obtained in an open-loop fashion : for some general design objective, such as maximum throughput of a manufacturing system or minimal heating along a spacecraft re-entry path, optimal trajectories are generated, assuming that the environment and initial conditions are fixed. This serves as an ideal reference but it cannot be expected that the plant will actually follow the optimized trajectory. For various reasons, including modelling errors, changes of the environment etc, deviations from the reference will occur and have to be compensated. This is achieved in a closed-loop fashion with the stabilization term: by feeding back some measure of the deviation, it is possible to stabilize the actual trajectory around the reference, so that the desired behaviour is obtained.

Despite its importance, the design of guidance laws is not discussed further in this chapter. Rather, attention is focused on the stabilization term, assuming that some reference trajectory has been obtained. As briefly discussed in chapter 1, this allows us to concentrate on linear hybrid dynamics, in fact linearized approximations around the desired trajectory of the original non linear plant state dynamics. This hypothesis will be discussed in chapter 7, as it is not so easily understood in the presence of regime uncertainty.

The objective of the present chapter is to propose a general approach to the design of the closed-loop part of the control law. For multivariable deterministic systems, the most popular approach in the state-space framework is to look for feedback gains that minimize an integral of quadratic penalizations in terms of deviations from the desired state and control levels. Because of its familiarity, this approach, called optimal LQ regulation (L for linear dynamics and Q for quadratic) is sometimes disconnected from its roots in the above guidance/stabilization dichotomy. It may therefore be worth stressing that the optimization of the quadratic functional in the LQ approach is only a tool and that there is most often no physical significance attached to the obtained minimum. In contrast the guidance optimization is based on physical objectives of the design.

It is in fact only in the most restricted situation that explicit optimization of the quadratic criterion is possible without limiting the class of admissible solutions. As in the deterministic setting for the LQ regulator, the optimal JLQ regulator (J for jump) to be obtained in § 3.2 is based on the assumption that a full measurement of the plant variables, here the state and regime, is available on-line through noiseless channels. In more realistic situations when only part of this information is available, constrained optimal regulators must be defined. To preserve tractability of the regulation laws, only optimization of linear feedback gains is considered here. The simplicity of this solution permits implementation of the regulator at a fast sampling rate to counter rapid changes in the plant environment. In § 3.3 we then measure how much optimality is lost, in terms of the quadratic functional. Various algorithms to compute the optimized gains are presented and tested in § 3.6.

Because of the stochastic nature of jump linear systems, the usual quadratic cost of LQ problems is now replaced (see equation (3.2.2) and its introduction below) by the mean value of a quadratic functional. The optimal JLQ regulator must therefore be understood as optimal only on average. This important point gives the incentive for the robustness analysis of chapter 4, where sample path value and distribution of the cost are considered.

3.2 The Jump Linear Quadratic Regulator

Throughout this chapter we limit attention to noiseless linear state dynamics, which, using the notations of chapter 1, are conveniently written as a piecewise deterministic differential equation

$$\dot{x}_t = A(r_t)\, x_t + B(r_t)\, u_t$$

where the regime r_t obeys a Markov chain. The regulation objective is to stabilize x_t around zero using a feedback control action u_t, so that the plant state stays near a nominal

x_n determined by the optimal guidance law. A nominal control u_n is associated to x_n and it is also desirable to limit the magnitude of u_t so that the plant control is close to u_n.

Let us first consider the simplest situation where both the state x_t and regime r_t of the hybrid model are measured. Denoting by $X_t = \sigma\text{-}\{r_s, t_0 \leq s \leq t\}$ and $R_t = \sigma\text{-}\{x_s, t_0 \leq s \leq t\}$ the corresponding σ-algebras, admissible control laws are restricted to $X_t \vee R_t$ measurable functions, but, as the pair (x_t, r_t) is Markov, it suffices to consider instantaneous feedback laws of the form

$$u_t = U(x_t, r_t, t)$$

with familiar smoothness conditions on U : $| U(x_1, r, t) - U(x_2, r, t) | \leq k | x_1 - x_2 |$ and $U(x, r, t) \leq k(1 + |x|)$ for some constant k (see appendix 2) . This class of admissible feedback laws is denoted by $\mathscr{U}$. For x_0 (= x_{t0}) some regime independent initial condition, the joint process (x_t, r_t) is Markov under any admissible control (it is however not true in general that x_t alone has the Markov property).

Of course this situation is not very realistic since we are often interested in regulating x_t around zero using a limited set of sensors, and x_t and r_t are in general not measured, but have to be estimated from the available sensor signals. The corresponding filtering problem is the subject of chapter 6 but later in this chapter constrained control structures are optimized with more realistic information requirements.

However, state and regime feedback is a good starting point in that it reveals many of the specificities of hybrid systems. Also it was the first problem studied historically, by Sworder (1969) and Wonham (1971).

As explained above the regulator task is to achieve small deviations from the nominal levels, as measured by the magnitudes of x_t and r_t. An approach which has had some success in a deterministic setting is to compute the regulator so as to minimize an integral of a quadratic function of x_t and u_t

$$\int_{t_0}^{t_f} (x_t' Q_1 x_t + u_t' Q_2 u_t)\, dt \qquad (3.2.1)$$

This form clearly captures part of the designer's objectives by giving a quadratic penalty to state and control excursions, but, as in the deterministic case, it is primarily chosen for mathematical reasons (existence, uniqueness of the solution).

The usual cost (3.2.1) cannot, however, be considered as a satisfactory performance index because, due to the stochastic nature of the system and to the jumps of r_t, it does not in general provide a complete ordering of competing control policies. The quantity (3.2.1) is in fact a random variable and it may happen that a control giving a good (small) value on one sample path gives a much larger one on another. This is a basic difficulty in designing optimal control laws for stochastic systems and it is classically by-passed by restricting attention to an average performance. This is typically illustrated by the LQG design where the problem is immediately transformed into a deterministic one with averaged control and estimation costs. A question to be analyzed is then the performance obtained on-line when implementing such "optimal on average" policies, but we defer this question until the next chapter and, for the present time, study only the simpler situation where (3.2.1) is replaced by its mathematical expectation

$$J = E \{ \int_{t_0}^{t_f} (x_t' \, Q_1 \, x_t + u_t' \, Q_2 \, u_t) \, dt \mid x_0, i_0, t_0 \} \tag{3.2.2}$$

conditionned upon the initial variables $x_{t_0} = x_0$, $r_{t_0} = i_0$. The symmetric matrices Q_1 and Q_2 satisfy the convexity conditions $Q_1 \geq 0$, $Q_2 > 0$ and, if needed, may be taken as regime dependent ($Q_1(r_t)$, $Q_2(r_t)$) to assign a different penalty to state and control excursions in different regimes. For a fault-tolerant control application, it may for example be decided to allow larger control excursions (i.e. take a "smaller" Q_2) after a failure, while in a nominal situation fuel economy may be favored.

The regulator design has now been completely formulated as an optimization problem

$$\min_{\mathcal{U}} J \qquad \text{subject to} \quad \dot{x}_t = A(r_t) \, x_t + B(r_t) \, u_t \tag{3.2.3}$$

and we present its solution using a stochastic version of the maximum principle and stochastic dynamic programming. In terms of the more general control objectives explained in § 3.1, it should be remembered that the optimization in (3.2.3), must be understood as a means to arrive at a systematic computation of the feedback part of the control law. The key practical asset of the resulting regulator will not be that it minimizes (3.2.3) but that it provides the desired stabilization around the reference trajectory. Since this most important aspect is not explicitly depicted in (3.2.3) it is necessary to study under what conditions the optimal design (3.2.3) actually produces a satisfactory stabilization law. This is the purpose of § 3.2.3.

3.2.1 Dynamic programming derivation

This derivation was proposed by (Wonham, 1971) using general results on the application of dynamic programming to stochastic optimization problems. The material required is reviewed in appendix 2.

The Jump Linear Quadratic (JLQ) regulator is given as

Theorem 3.1

The solution of (3.2.3) is

$$u_t^* = -Q_{2i}^{-1} B_i \Lambda_i x_t \qquad \text{when } r_t = i \tag{3.2.4}$$

with the Λ_is, i = 1 to N, obtained from a set of coupled Riccati equations

$$- \dot{\Lambda}_i = A_i' \Lambda_i + \Lambda_i A_i - \Lambda_i B_i Q_{2i}^{-1} B_i' \Lambda_i + Q_{1i} + \sum_{j=1}^{N} \pi_{ij} \Lambda_j \qquad ; i = 1, N \tag{3.2.5}$$

$$\Lambda_i (t_f) = 0$$

and the minimal cost is

$$J^* = x_0' \Lambda_{i0} (t_0) x_0$$

for $x_{t0} = x_0$ and $r_{t0} = i_0$.

Proof :

Introducing the conditional cost-to-go as

$$V(x, i, t) = \min_{\mathscr{U}} E \{ \int_t^{t_f} (x_s' Q_1 x_s + u_s' Q_2 u_s) ds \; | x_t = x, r_t = i, t\}$$

the verification theorem of dynamic programming leads to

$$0 = \min_{u \in \mathbf{R}^m} \{V_t (x, i, t) + \mathscr{L}_u V(x, i, t) + x_t' Q_1 x_t + u_t' Q_2 u_t\} \tag{3.2.6}$$

for $t_0 \leq t \leq t_f$, $(x, i) \in \mathbf{R}^n \times \mathbf{S}$

with the boundary condition

$$V(x, i, t_f) = 0$$

where $\mathscr{L}_u$ is the generator of the Markov pair (x_t, r_t) under an admissible action u. The generator is computed as in chapter 2

$$\mathscr{L}_u V(x, i, t) = (A_i x + B_i u)' V_x (x, i, t) + \sum_{j=1}^{N} \pi_{ij} V(x, j, t)$$

The optimization problem has thus been reduced to two sub-problems: solve the second order partial differential equation (3.2.6), and express the optimal control u^* in terms of the corresponding minimal cost-to-go. This second step is easy because the minimization in (3.2.6) can be performed explicitly to find

$$u_t = - Q_{2i}^{-1} B_i' V_x (x, i, t)$$

Bringing u_t back into (3.2.6), an equation in V is obtained which is solved by guessing a quadratic form in x for the solution

$$V (x, i, t) = x' \Lambda_i x$$

where the time dependence of Λ_i is not explicited. At the minimum in (3.2.6) we obtain

$$0 = x'(\dot{\Lambda}_i + A_i' \Lambda_i + \Lambda_i A_i - \Lambda_i B_i Q_{2i}^{-1} B_i' \Lambda_i + Q_{1i} + \sum_{j=1}^{N} \pi_{ij} \Lambda_j)x$$

which is true for any (x, i), hence giving (3.2.5). At the boundary we obtain

$$V(x, i, t_f) = x' \Lambda_i (t_f) x = 0 \qquad \text{for any } (x, i)$$

so that the boundary conditions for (3.2.5) are $\Lambda_i (t_f) = 0$, i = 1, N. Finally going back to the cost-to-go expressed at time $t = t_0$, the minimal cost is

$$J^* = V(x_0, i_0, t_0) = x_0' \Lambda_{i0} (t_0) x_0$$

which completes the proof. □

It is important that the solution of the JLQ regulator has been obtained as a feedback law, and more precisely as a linear regime dependent state feedback

$$u_t^* = \Gamma_i x_t \qquad \text{when } r_t = i$$
$$\text{with } \Gamma_i = - Q_{2i}^{-1} B_i' \Lambda_i$$

This is very similar to the familiar LQ situation with the added complexity of a regime dependent gain and of the coupling ($\sum \pi_{ij} \Lambda_j$) for the corresponding Riccati equations. Of course as the jumps become more and more rare (π_{ij}, i, j = 1, N, small), the set (3.2.5) reduces to the set of N decoupled Riccati equations associated with N LQ regulators for regime r = 1 to N. However, the coupled equations (3.2.5) are to be solved off-line (some algorithms are given in § 3.6 below) and on-line implementation requires only a few real-time multiplications (m x n) and additions (m x n).

The structure of the JLQ regulator is interesting with regard to the applications mentioned in chapter 1. Theorem 3.1 states that the optimal way of using x_t and r_t is to feed them back in a control law $u_t = \Gamma_i x$ when $r_t = i$, as shown in figure 3.1.

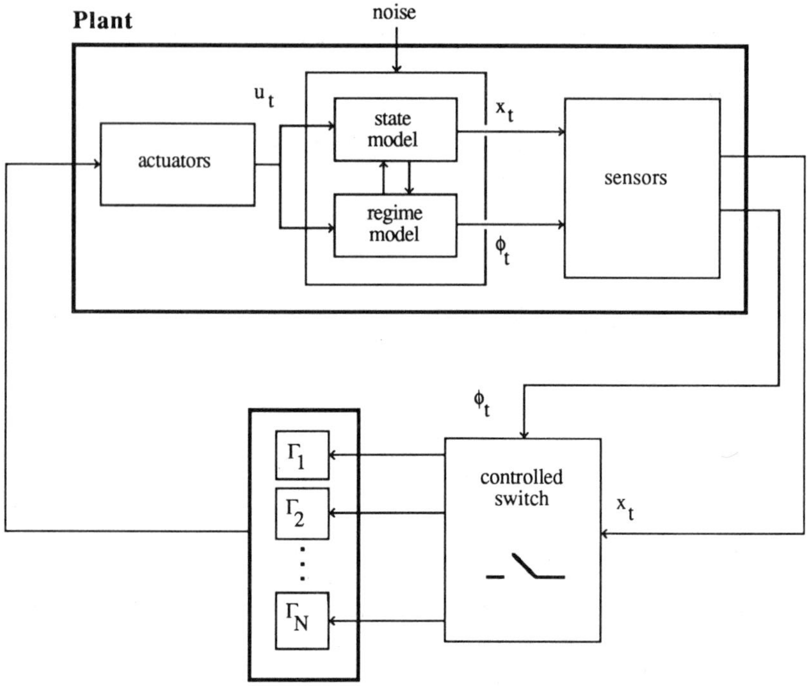

Figure 3.1 - Structure of the JLQ regulator.

For the fault-tolerant control application this provides the desired reconfiguration : once a failure occurs, that is once r_t jumps, say from i to j, the optimal adaptation is to switch the gain from Γ_i to Γ_j. Similarly when the insolation over the field of heliostats of the solar thermal receiver drops due to the passage of a cloud, the loop gains are optimally adapted by (3.2.4). This also points out the crucial way (3.2.4) uses the assumption of state and regime observation, but its structure can nevertheless serve as a reference in more realistic situations where no such sensors are available.

3.2.2 Maximum principle derivation

This is the derivation of theorem 3.1 originally presented by Sworder (1969). A summary of the application of the maximum principle to stochastic optimization is given in appendix 2.

For $u = U(x, r, t)$ an admissible control law, the stochastic hamiltonian of the system is formed as

$$H(x, U, t) = \lambda_t' (A(r_t)\, x_t + B(r_t)\, u_t) + x_t'\, Q_1\, x_t + u_t'\, Q_2\, u_t$$

where $\lambda_t \in R^n$ is the co-state or adjoint vector, determined by the equation

$$-\dot{\lambda}_t = (\partial H/\partial x)' \qquad \lambda(t_f) = 0 \tag{3.2.7}$$

Note that the cost integrand appears with a positive sign in the above hamiltonian so that we indeed use the minimum principle rather than the maximum principle presented in appendix 2. Of course the two principles are equivalent up to sign change.

From appendix 2, the optimal law U^* minimizes the conditional expectation of the hamiltonian

$$E\,\{H(x, U, t)\,|x_t, r_t\} \;\geq\; E\,\{H(x, U^*, t)\,|x_t, r_t\} \quad \text{a.s.}$$

But there are no constraints and we solve directly

$$u_t^* = -\frac{1}{2} Q_2^{-1}\, B'(r_t)\, E\,\{\lambda_t\, |x_t, r_t\}$$

To determine the solution, assume that the co-state takes the form

$$\lambda_t = 2\Lambda\, x_t \qquad t_0 \leq t \leq t_f$$

where Λ is a random symmetric matrix, conditionally independent on x and differentiable everywhere. Then

$$u_t^* = -\, Q_2^{-1}\, B'(r_t)\, E\,\{\Lambda | r_t\}\, x_t \tag{3.2.8}$$

With the notation $\Lambda_i = E\,\{\Lambda\, \phi_{ti}\}$, (3.2.8) yields the optimal control as in theorem 3.1

$$u_t^* = -\, Q_2^{-1}\, B_i'\, \Lambda_i\, x_t \qquad \text{when } r_t = i$$

We carry this expression back into the co-state equation (3.2.7) and find that

$$
\begin{aligned}
-\dot{\Lambda} = & A(r_t)'\Lambda + \Lambda A(r_t) - \Lambda B(r_t)\, Q_2^{-1}\, B(r_t)'\, E\,\{\Lambda | r_t\} \\
& - E\,\{\Lambda | r_t\}\, B(r_t)\, Q_2^{-1}\, B(r_t)'\, \Lambda \\
& + E\,\{\Lambda | r_t\}\, B(r_t)\, Q_2^{-1}\, B(r_t)'\, E\,\{\Lambda | r_t\} + Q_1
\end{aligned}
\tag{3.2.9}
$$

Equation (3.2.9) is a stochastic differential equation in Λ, integrated backwards in time from the boundary condition $\Lambda(t_f) = 0$ (free terminal state, $\lambda(t_f) = 0$). Fortunately it is actually the set of matrices Λ_i , $i = 1$ to N, that is needed to compute u_t^* as r_t varies and it is possible to deduce from (3.2.9) a set of deterministic equations for the Λ_is.

From the definition of Λ_i we have

$$d\Lambda_i = E\,\{d\Lambda\, \phi_{ti}\} + E\,\{\Lambda\, d\phi_{ti}\}$$

where variables are propagated backwards. For Π' the forward generator of the regime Markov chain, the second expectation yields

$$-\sum_{j=1}^{N} \pi_{ij}\, \Lambda_j\, dt$$

We multiply (3.2.9) by ϕ_{ti} dt and the first expectation is computed as

$$-(A_i'\,\Lambda_i + \Lambda_i\, A_i - \Lambda_i B_i\, Q_{2i}^{-1}\, B_i'\, \Lambda_i + Q_{1i})dt$$

so that

$$d\Lambda_i = -(A_i'\,\Lambda_i + \Lambda_i\, A_i - \Lambda_i B_i\, Q_{2i}^{-1}\, B_i'\, \Lambda_i + Q_{1i})dt - \sum_{j=1}^{N} \pi_{ij}\, \Lambda_j\, dt$$

and, dividing by dt, the set of coupled Riccati equations is again obtained. This completes the maximum principle derivation of theorem 3.1.

As presented here, it is apparent that there is no advantage in using one rather than the other way to derive theorem 3.1. As explained in appendix 2, the stochastic maximum principle encompasses a larger class of optimization problems by relaxing the smoothness conditions imposed on the cost integrand when using dynamic programming. For the quadratic penalties considered here, these conditions are trivially satisfied but they might be more restrictive in other situations.

It is also interesting to note that the co-state equation is formally identical to its counterpart for the LQ deterministic problem. From (3.2.7) we compute

$$-\dot{\lambda}_t = A(r_t)'\,\lambda_t + Q_1\,x_t$$

which is the dual of the state equation

$$\dot{x}_t = A(r_t)\,x_t + B(r_t)\,u_t$$

The familiar two point boundary value problem of the maximum principle is thus recovered as two piecewise deterministic differential equations coupled by the control extremality condition $u_t = u_t^*$. Iterating off-line, forward ($\dot{x}_t$) and backward ($\dot{\lambda}_t$) integrations of these equations for a known sample path of r_t over $[t_0, t_f]$ provides an open-loop solution of the optimization problem. In practice it is of course the closed-loop solution that is valuable.

3.2.3 Asymptotic behaviour

Theorem 3.1 provides a regulator with time-varying gains (3.2.4) and, from the point of view of implementation, it would be interesting to simplify the solution by using stationary expressions of the gains. More importantly, the stabilizing properties of this regulator need to be assessed.

It is well known in the deterministic LQ context (Anderson and Moore 1971, Kwakernaak and Sivan 1972) that these questions are related to the asymptotic behavior of the solution given by theorem 3.1. Intuition suggests that as $t_f - t_0 \rightarrow \infty$, and provided that the matrices A_i, B_i, Q_{1i} and Q_{2i} do not depend on time, the set of Riccati equations

(3.2.5) should approach its stationary counterpart ($\dot{\Lambda}_i = 0$, i = 1, N). Thus, possibly under some technical conditions, the corresponding closed-loop system should be stable. For the LQ problem it was found that the additional conditions needed are stabilizability (there must exist a control that stabilizes the system) and detectability (the unstable part of the state trajectory must be reflected in the cost function).

For the present JLQ problem the situation is more involved. We may first try to replicate the conditions of the asymptotic LQ regulator for each regime and then to constrain the jump process so that these conditions also work in JLQ case. This is the approach of (Wonham, 1971) and, essentially, it views the regime transitions as perturbations to an underlying deterministic problem. The result is then a conservative sufficient condition in term of "rare jumps". This inconvenience might be overlooked in situations where we want to characterize the controlled system independently in each regime: the stability obtained is the usual deterministic asymptotic stability for modified regime dynamics considered separately.

On the other hand, we may also try to obtain sharper necessary and sufficient conditions by considering the stochastic nature of the hybrid system. Following the LQ case, the stochastic notions of stabilizability and detectability introduced in chapter 2 should be relevant to this discussion as shown by theorem 3.3 below. From the examples presented in chapter 2 it is then apparent that the corresponding asymptotic stability condition cannot provide a regime-by-regime result but rather a global characterization of the averaged dynamics involving the interaction of discrete and continuous components.

The two approaches encompass different aspects of the asymptotic JLQ regulator and are therefore presented together, starting with Wonham's result.

Theorem 3.2

Suppose that the triplets $[A_i, B_i, Q_{1i}^{1/2}]$, i = 1 to N, are stabilizable and detectable in the usual deterministic sense, and that the regime generator is such that

$$\max_{i=1,N} \inf_{\Gamma} \left| \pi_{ii} \int_0^{\infty} \exp\left[\left(A_i + B_i + \frac{1}{2}\pi_{ii} I_n\right)'t\right] \exp\left[\left(A_i + B_i + \frac{1}{2}\pi_{ii} I_n\right)t\right] dt \right| < 1 \quad (3.2.10)$$

The Λ_i, i = 1 to N, solutions of (3.2.5) then have the following properties

(i) The limits $\bar{\Lambda}_i = \lim_{t_0 \to -\infty} \Lambda_i$ exist and are the unique positive semi-definite solutions of a set of coupled algebraic Riccati equations

$$0 = A_i'\bar{\Lambda}_i + \bar{\Lambda}_i A_i - \bar{\Lambda}_i B_i Q_{2i}^{-1} B_i' \bar{\Lambda}_i + Q_{1i} + \sum_{j=1}^{N} \pi_{ij} \bar{\Lambda}_j \qquad ; i = 1, N \qquad (3.2.11)$$

(ii) The matrices $A_i - B_i Q_{2i}^{-1} B_i' \bar{\Lambda}_i + \frac{1}{2}\pi_{ii} I_n$, i = 1 to N, are stable in a deterministic sense.

<u>Proof</u> :

The original proof was given in (Wonham, 1971), the key step is as follows. For some constant gains Γ_i such that $A_i + B_i\Gamma_i$ is stable (for i = 1, N) and for the algebraic counterpart of (3.2.5) (that is (3.2.5) with $\dot{\Lambda}_i = 0$, i = 1, N) transformations T_i, i = 1, N, of the Λ_is are defined as

$$T_i(\Lambda_1, \Lambda_2, \ldots, \Lambda_n) = \int_0^{\infty} [\exp{(A_i + B_i\Gamma_i + \tfrac{1}{2}\pi_{ii} I_n)'t}]$$

$$[\sum_{\substack{j=1 \\ j\neq i}}^{N} \pi_{ij} \Lambda_j][\exp{(A_i + B_i\Gamma_i + \tfrac{1}{2}\pi_{ii} I_n)t}] \; dt$$

Convergence of the solutions (3.2.5) to the solutions (3.2.11) when $t_0 \to -\infty$ is then proved by using (3.2.10) to guarantee that the T_is are contractive transformations. □

Condition (3.2.10) is crucial to this result and its conservativeness is difficult to evaluate. It favors systems with rare jumps, and this can be expected after the selection of a deterministic stabilizability and detectability condition for the triplets $[A_i, B_i, Q_{1i}^{1/2}]$, i = 1, N. Indeed in the limit when $\pi_{ii} \to 0$, i.e. the no jump case, theorem 3.2 reduces to the asymptotic LQ result separately replicated N times. Condition (3.2.10) for a scalar system with $B_i = 0$ and $A_i = 2$ is satisfied for $|\pi_{ii}| = 1$ (the left hand side is 1/3) and violated for $|\pi_{ii}| = 8$ (the left-hand side is 4). Jumps separated on the average by one hour can be tolerated when jumps occuring on the average every eighth of hour are not permitted. The restriction is compensated by the fact that theorem 3.2 produces a regime by regime deterministic stability condition. Observe, however, that $A_i - B_i Q_{2i}^{-1} B_i' \bar{\Lambda}_i + \frac{1}{2}\pi_{ii} I_n$ is stabilized but not the closed-loop regime dynamics $\tilde{A}_i = A_i - B_i Q_{2i}^{-1} B_i' \bar{\Lambda}_i$. Because

π_{ii} is negative, instability of $\tilde{A}_i$ is therefore not ruled out and this phenomenon is in a sense the last influence of the stochastic nature of the system left in the deterministic approach of theorem 3.2. It is indeed reminiscent of chapter 2 where the influence of a short average life time in the tolerance of unstable regimes was demonstrated. The occurrence of this instability is further illustrated by the following example.

Example

Consider a scalar system with two regimes, $a_1 = -1$, $a_2 = 1$, $b_1 = b_2 = 1$ and a transition rate matrix $\Pi = \begin{pmatrix} 0 & 0 \\ 2\pi & -2\pi \end{pmatrix}$. Regime 1 is absorbing and regime 2 transient with an average life time of $1/2\pi$ hours. With unity weighting coefficients, the asymptotic JLQ regulator of Theorem 3.2 produces for regime 2

$$\tilde{a}_2 = \pi - \sqrt{(\pi - 1)^2 + 1 + 2\pi(\sqrt{2} - 1)}$$

which is unstable for π greater than $1/(2-\sqrt{2})$ while the shifted closed-loop dynamics,

$$\tilde{a}_2 + \frac{1}{2}\pi_{22} = -\sqrt{(\pi - 1)^2 + 1 + 2\pi(\sqrt{2} - 1)}$$

are unconditionally stable.

Theorem 3.2 nevertheless provides a deterministic type of stability for the shifted dynamics $\tilde{A}_i + \frac{1}{2}\pi_{ii}I_n$. In practice it is sometimes desirable to achieve deterministic stability of $\tilde{A}_i$ itself and a modification of theorem 3.2 is then needed where the "stabilizing" shift $\frac{1}{2}\pi_{ii}I_n$ would be counterbalanced. This point is further discussed in connection with the robustness analysis of JLQ regulators in chapter 4 (see in particular § 4.6).

The treatment of regime transitions as small perturbations of a set of otherwise deterministic dynamics is not appropriate to situations such as the target tracking application in chapter 1, where jumps occur with an average frequency well within the system bandwidth. Another approach is then needed which better reflects the stochastic nature of the hybrid model. Among the various notions of stochastic stability reviewed in appendix 2, we focus on asymptotic mean square stability. Given the quadratic form of the performance index, it seems reasonable to expect that the optimal regulator guarantees this type of stability, possibly under certain additional assumptions. At this point the reader is referred back to chapter 2, §2.4, where Liapunov's methods were used to obtain mean square stabilizability conditions. It is recalled here that this notion of stabilizability globally characterizes the pairs $[A_i, B_i]$, $i = 1, N$, as they are coupled by regime jumps with rates $(\pi_{ij})_{i,\, j\, =\, 1,\, N}$, so that one should speak of the stabilizability of the $[A_i, B_i, i = 1, N, \Pi]$ model. The dual notion of stochastic detectability for $[A_i, C_i, i = 1, N, \Pi]$ is also useful here (see chapter 2, § 2.5).

The idea is simply that, with the appropriate notions of stochastic stabilizability and detectability, conditions for the asymptotic JLQ regulator mimic the well-known conditions of the deterministic LQ problem (e.g. Kwakernaak and Sivan 1972, Theorem 3.7). The result is

<u>Theorem 3.3</u>

The limits $\bar{\Lambda}_i = \lim\limits_{t_0 \to -\infty} \Lambda_i$, $i = 1, N$ exist as the unique positive semi-definite solutions of (3.2.11) and the closed-loop dynamics

$$\dot{x}_t = [A(r_t) - B(r_t)Q_2^{-1}\, B'(r_t)\, \bar{\Lambda}(r_t)]x_t$$

are then exponentially stable in mean square if and only if $[A_i, B_i, i = 1, N, \Pi]$ and$[A_i, Q_{1i}^{-1/2}, i = 1, N, \Pi]$ are, respectively, stochastically stabilizable and detectable.

<u>Proof</u> :

Necessity :

(i) If the system is not detectable, then for any feedback gains the performance index is unbounded on $t_0 < t_f < \infty$ and the minimization cannot stabilize the system. Detectability is therefore necessary.

(ii) If the system is not stabilizable, no control law, and hence in particular not the JLQ regulator, can stabilize the system. Stabilizability is therefore necessary.

Sufficiency :

If the system is stabilizable, then the gains Γ_i , i = 1 to N, of § 2.4.1 lead to a finite performance which constitute an upper-bound for the cost-to-go

$$J_t = E\{ \int_t^{t_f} (x_\theta' Q_1 x_\theta + u_\theta' Q_2 u_\theta)\, d\theta | x_t, r_t, t\}$$

But this cost is clearly a monotone non-decreasing function of t_f and therefore it converges. Its value is related to the Λ_is by

$$J_t = x_t' \Lambda_i x_t \qquad \text{when } r_t = i$$

so that the convergence of J_t implies that of Λ_i , i = 1 to N, say towards $\bar{\Lambda}_i$, i = 1 to N. From the continuity of the solutions of the (3.2.5) it is then clear that the $\bar{\Lambda}_i$s have to satisfy (3.2.11). If the system is detectable, then the gains Γ_i , i = 1 to N, of the detectability theorem in chapter 2 guarantee that the integral

$$P = E\{ \int_{t_0}^{t} (\exp \tilde{A}' (\theta - t_0)\, Q_1 \exp \tilde{A} (\theta - t_0))\, d\theta | x_0, i_0, t_0\}$$

is bounded on $t_0 < t < \infty$ by the P_is, the corresponding Liapunov matrices of theorem 2.6. The quadratic function

$$V(x, \phi, t) = x' P x$$

then qualifies as a Liapunov function to ensure that the closed-loop system is exponentially stable in mean square (see § 2.4.1). □

The main interest of this theorem is that the conditions obtained are necessary and sufficient. It is clear, however, that they are not so easily tested and we would certainly accept giving up necessity to simplify the conditions in order to arrive at an algebraic rank test as in the LQ setting. The discussion in chapter 2 on the relationship between stochastic controllability and stochastic stabilizability should be developed to arrive at new results in this direction. In the class of hybrid systems with directed chains (see appendix 1) it can be shown (Ji and Chizeck, 1987) that deterministic controllability and observability for the pairs $[A_i, B_i]$, $[A_i, Q_i^{-1/2}]$, i = 1, N, is sufficient for the asymptotic JLQ regulator to converge toward a stabilizing (in the mean square sense) control law.

As the stability obtained from theorem 3.3 is of a stochastic type, averaging of state dynamics through regime jumps occurs and individual regimes may remain unstable provided their average life time is short enough.

Finally we illustrate by an example the conservativeness of theorem 3.2

<u>Example</u> :

Consider a system in $\mathbf{R}^2$ with two modes and

$$A_1 = \begin{bmatrix} 2 & 1 \\ -7 & -3 \end{bmatrix} \qquad B_1 = 0$$
$$A_2 = \begin{bmatrix} -3 & -1 \\ 7 & 2 \end{bmatrix} \qquad B_2 = 0$$

and a matrix of transition rates $\Pi = \begin{bmatrix} -4 & 4 \\ 4 & -4 \end{bmatrix}$.

The control matrices B_1 and B_2 being null, the two regimes are deterministically stabilizable only if they are stable. This is true since the eigenvalues of A_1 and A_2 are the solutions of

$$\lambda^2 + \lambda + 1 = 0$$

However, the rare jumps condition of theorem 3.2 is not satisfied because the left-hand side of (3.2.10) is equal to 4.64 (> 1). By computing the Liapunov stability condition as explained in chapter 2, it is found that the averaged dynamic matrix is $A_1 + A_2$ which is stable $(A_1 + A_2 = -I)$. If the cost is chosen such that stochastic detectability holds, this example illustrates a violation of the conditions of theorem 3.2 while the necessary and sufficient conditions of theorem 3.3 are satisfied.

3.3 Suboptimal Solutions

The optimal JLQ regulator provides a rational way of selecting the control gains when both the state and regime are measured. However feedback laws using a more limited set of sensors are often implemented and subsequent sections consider the problem of optimizing the corresponding structurally constrained regulators. Here we show how competing solutions may be compared to the JLQ regulator in terms of a suboptimality degree. The full state and regime feedback solution, though unrealistic, then appears as a valuable reference, to be computed at an early stage of the design study, in order to be able to rank more realistic structures and to analyze trade-offs.

As an arbitrary control law take

$$u_t = \Gamma_i x_t \qquad \text{when } r_t = i \tag{3.3.1}$$

for some gains Γ_i, i = 1, N. The non-switching, or regime independent solutions are included in (3.3.1) via $\Gamma_i = \Gamma$, i = 1, N and output feedback schemes can be obtained by choosing the Γ_is as $\Gamma_i = \bar{\Gamma}_i C_i$, i = 1, N. The value of the cost (3.2.2) under (3.3.1) serves as a measure of the quality of the gains Γ_i, i = 1, N : if they were chosen as in (3.2.4) then J would attain its minimal value J^* and therefore the closer J comes to its minimum the better the gains. This is measured by the suboptimal degree υ, defined as

$$\upsilon = \frac{J - J^*}{J^*}$$

which gives the relative degradation of the cost for using a non optimal control law. Recalling that J^* was given in theorem 3.1 as $J^* = x_0' \Lambda_{i_0}(t_0)\, x_0$, what remains to be computed is J, the value of (3.2.2) under the control (3.3.1). The result is given by the following theorem

Theorem 3.4

Under the arbitrary control law (3.3.1) the value of the cost is

$$J = x_0' P_{i_0}(t_0)\, x_0 \qquad \text{when } x(t_0) = x_0,\ r(t_0) = i_0 \tag{3.3.2}$$

where the P_is are solutions of a set of coupled Liapunov equations

$$- \dot{P}_i = \tilde{A}_i' P_i + P_i \tilde{A}_i + Q_{1i} + \Gamma_i' Q_{2i} \Gamma_i + \sum_{j=1}^{N} \pi_{ij} P_j$$

$$P_i (t_f) = 0 \tag{3.3.3}$$

with $\tilde{A}_i = A_i + B_i\Gamma_i$, i = 1, N.

Proof :

From the cost definition we obtain

$$J = E \{x_{t_0}' [\int_{t_0}^{t_f} \exp \tilde{A}'(t - t_0) (Q_1 + \Gamma' Q_2 \Gamma) \exp \tilde{A} (t - t_0)) dt] x_{t_0} | x_0, i_0, t_0\} \tag{3.3.4}$$

This is a quadratic form in $x_{t_0} = x_0$ and the corresponding matrix is introduced

$$P (t) = \int_{t}^{t_f} (\exp \tilde{A}' (\tau - t_0) (Q_1 + \Gamma' Q_2 \Gamma) \exp \tilde{A} (\tau - t_0)) d\tau$$

In this definition dependence on r_t has not been explicited but it is clearly present through $\tilde{A} = A(r_t) + B(r_t) \Gamma(r_t)$, and, if desired by the designer, $Q_1 = Q_1(r_t)$, $Q_2 = Q_2(r_t)$. The matrix P(t), defined by the above stochastic integral, is also the solution of a piecewise deterministic differential equation

$$\dot{P} = -\tilde{A}'P - P\tilde{A} - Q_1 - \Gamma' Q_2 \Gamma$$

as can be seen by direct differentiation with respect to time. The boundary condition for this equation is immediately deduced from the integral definition of P as $P(t_f) = 0$. Using again the notation $P_i = E \{P \phi_{ti}\}$, i = 1, N, the suboptimal cost is, from (3.3.4),

$$J = x_0' P_{i_0} (t_0) x_0$$

The P_is are obtained by noting that

$$dP_i = E\ \{dP\ \phi_{ti}\} + \{P\ d\phi_{ti}\}$$

where the variables are propagated backwards. But multiplying $\dot{P}$ by ϕ_{ti} dt and taking the expectation gives

$$E\ \{dP\ \phi_{ti}\} = (\ - \tilde{A}_i' P_i - P_i \tilde{A}_i - Q_{1i} - \Gamma_i' Q_{2i}\, \Gamma_i)\ dt$$

and for the Markov chain model of ϕ_t

$$E\ \{dP\ \phi_{ti}\} = \sum_{j=1}^{N} \pi_{ij}\ P_j\ \ dt$$

With these last two equations in dP_i (3.3.3) has been proved, the boundary condition $P(t_f) = 0$ being translated as $P_i(t_f) = 0$, i = 1, N. □

As in the LQ situation, the relationship between Riccati and Liapunov equations is observed, the former characterizing optimal solutions and the latter being associated with suboptimal solutions and stability conditions (see chapter 2). It can be observed that the (optimal) gains of theorem 3.1 ($\Gamma_i = -Q_{2i}^{-1}\ B_i'\ \Lambda_i$) in (3.3.3) transform the coupled Liapunov equations into Riccati equations. In fact starting with (3.3.3) this could serve as a purely algebraic proof of the optimality conditions of theorem 3.1.

The suboptimality degree υ can thus be computed for any linear feedback law and different gains can be compared. We recall, however, that the comparison based on υ cannot encompass all aspects of a realistic design, since it is solely based on the simple mathematical index J which does not explicitly capture important issues like regulator complexity or robustness margins. Within the limits of quadratic performance evaluation the usefulness of theorem 3.4 is illustrated in § 3.6 below.

3.4 The Optimal Switching Output Feedback

It is assumed here that the regime is observed but that only part of the state is measured through a regime dependent and noise-free channel

$$y_t = C(r_t)\, x_t \tag{3.4.1}$$

The dependence of C on the regime would be useful in the fault-tolerant control application to describe sensor failures. Even though x_t is not completely observed, there is no polluting noise in (3.4.1) which for the time being allows for a direct feedback of the sensor outputs. The more general situation where there is some measurement noise and where filtering of measurements is required is considered in chapter 5.

An easily implemented regulator using r_t and y_t is

$$u_t = \Gamma_i\, y_t = \Gamma_i\, C_i\, x_t \qquad \text{when } r_t = i \tag{3.4.2}$$

and we want to find the optimal gains for this constrained structure, that is to solve the optimization problem

$$\min_{\Gamma_i,\, i = 1,N} J \qquad \text{subject to } \dot{x}_t = A(r_t)\, x_t + B(r_t)\, u_t \quad \text{and } (3.4.1), (3.4.2). \tag{3.4.3}$$

The answer is found by transforming (3.4.3) into an equivalent deterministic problem, averaged over the initial state. For that purpose it is assumed that x_0 is a centered random variable with E $\{x_0 x_0' \, \phi_{t_0 i}\} = X_{i0}$ where ϕ_{t_0} is the regime indicator at t = t_0.

<u>Theorem 3.5</u>

A necessary condition for (3.4.3) to be optimal is that the gains Γ_i, i = 1,N satisfy

$$\dot{X}_i = \tilde{A}_i X_i + X_i \tilde{A}_i' + \sum_{j=1}^{N} \pi_{ji} X_j \qquad ; \; X_i(t_0) = X_{i0} \tag{3.4.4.a}$$

$$-\dot{\Lambda}_i = \tilde{A}_i' \Lambda_i + \Lambda_i \tilde{A}_i + Q_{1i} + C_i' \Gamma_i' Q_{2i} \Gamma_i C_i + \sum_{j=1}^{N} \pi_{ij} \Lambda_j \qquad ; \; \Lambda_i(t_f) = 0 \tag{3.4.4.b}$$

$$\Gamma_i = -\, Q_{2i}^{-1} B_i' \Lambda_i X_i C_i' (C_i X_i C_i')^{-1} \tag{3.4.4.c}$$

where $\tilde{A}_i = A_i + B_i \Gamma_i C_i \quad , i = 1, N.$

Proof :

Averaging over the initial state and permuting integration and expectation (which is allowed for a sufficiently smooth integrand by Fubini's theorem, see e.g. (Rudin, 1966, p. 140)) the cost under the feedback law (3.4.2) can be written

$$\int_{t_0}^{t_f} \text{trace } E\, \{x_t\, x_t' \,(Q_1 + C'\,\Gamma'\, Q_2\, \Gamma\, C) \mid x_0, i_0, t_0\}dt$$

Next N matrices X_i (n x n) are defined as $X_i = E\,\{x_t\, x_t'\, \phi_{ti}\}$, i = 1 to N. To propagate (forward in time) the X_is note that

$$dX_i = E\,\{d(x_t\, x_t')\, \phi_{ti}\} + E\,\{x_t\, x_t'\, d\phi_{ti}\}$$

and

$$d(x_t\, x_t') = (\tilde{A}\, x_t\, x_t' + x_t\, x_t'\, \tilde{A}')dt$$

where $\tilde{A}(r_t) = A(r_t) + B(r_t)\ \Gamma(r_t)\, C(r_t)$. The second expectation gives

$$E\,\{x_t\, x_t'\, d\phi_{ti}\} = \sum_{j=1}^{N} \pi_{ji}\, X_j\ \ dt$$

and the first one

$$E\,\{d(x_t\, x_t')\, \phi_{ti}\} = (\tilde{A}_i\, X_i + X_i\, \tilde{A}'_i)dt$$

so that (3.4.4.a) is obtained. Using the X_is, (3.4.3) is transformed into an equivalent (in an average sense) deterministic problem

$$\underset{\Gamma_i,\ i=1,N}{\text{Min }} J \quad \sum_{i=1}^{N} \int_{t_0}^{t_f} \text{trace } X_i\, (Q_{1i} + C_i'\, \Gamma_i'\, Q_{2i}\, \Gamma_i\, C_i)\, dt \tag{3.4.5}$$

subject to $\dot{X}_i = \tilde{A}_i\, X_i + X_i\, \tilde{A}_i' + \sum_{j=1}^{N} \pi_{ji}\, X_j \quad ; \ X_i\,(t_0) = X_{i0}$

For this problem the matrix maximum principle (see appendix 2) is useful to obtain simply necessary conditions for optimality. The hamiltonian associated to (3.4.5) is

$$H = \sum_{i=1}^{N} \text{trace } X_i \,(Q_{1i} + C_i' L_i' Q_{2i} L_i C_i) + \text{trace } (\tilde{A}_i X_i + X_i \tilde{A}_i' + \sum_{j=1}^{N} \pi_{ji} X_j)\Lambda_i$$

where the Λ_i , i = 1 to N, are n x n adjoint (or co-state) matrices. The optimality conditions are then

$$\begin{aligned} \partial H/\partial \Lambda_i &= \dot{X}_i \\ \partial H/\partial X_i &= -\dot{\Lambda}_i \qquad ; i = 1 \text{ to } N \\ \partial H/\partial \Gamma_i &= 0 \end{aligned}$$

which are easily computed (see the list of derivatives w.r.t. matrices of appendix 2) to give (3.4.4.a, b and c). □

These optimality conditions define a two-point boundary-value problem (TPBVP) : the state matrix equations (3.4.4.a) are integrated forward in time with $X_i(t_0) = X_{i0}$, i = 1 to N, and the co-state matrix equations (3.4.4.b) are integrated backward from $\Lambda_i(t_f) = 0$, i = 1 to N. The expressions (3.4.4.c) for the optimal gains couple the two integrations in a nonlinear manner and the numerical solution of (3.4.4) is rather involved. It can be noted once more that the duality between state and co-state is also present through the respective coupling intensities, π_{ij} for the i-th backward equation and π_{ji} for the i-th forward one.

To arrive at a simple feedback solution in theorem 3.5, the cost was averaged over the initial state x_0, assuming a centered initial condition with known covariance. In practice the choice of X_{i0} is not straightforward ; $X_{i0} = I_n$, corresponding to a uniform distribution of x_0 in $\mathbf{R}^n$, is often proposed, probably because it looks like an unprejudiced choice, but it is rarely motivated from physical considerations. The fact is that in many applications some combinations of state components are known to be unlikely at the initial time while others correspond to usual initial conditions (equilibrium points, etc). It is then advisable to give more emphasis (in the average cost) to these conditions that are expected from a physical point of view. Carrying this idea one step further an alternative for the design of constrained regulators can be proposed. After a preliminary computation of the optimal unconstrained solution, the constrained solution is designed to preserve, for example, the eigenspace of the optimal closed-loop dynamics corresponding to initial conditions. This is called a projective solution and is quite well-known in a deterministic LQ setting (see § 3.7 below).

To alleviate the difficulty of the nonlinear TPBVP (3.4.4) the asymptotic ($t_f \to \infty$) regulator may be sought for. The precise conditions under which this limit can be taken are not detailed here (see the discussion and references in § 3.7). When valid, the counterpart of (3.4.4) is

$$0 = \tilde{A}_i X_i + X_i \tilde{A}'_i + \sum_{j=1}^{N} \pi_{ji} X_j + X_{i0}$$

$$0 = \tilde{A}'_i \Lambda_i + \Lambda_i \tilde{A}_i + Q_{1i} + C'_i \Gamma'_i Q_{2i} \Gamma_i C_i + \sum_{j=1}^{N} \pi_{ij} \Lambda_j$$

$$\Gamma_i = - Q_{2i}^{-1} B'_i \Lambda_i X_i C'_i (C_i X_i C'_i)^{-1}$$

with $\tilde{A}_i = A_i + B_i \Gamma_i C_i \quad , i = 1, N$

An important related question in then the stability of the resulting closed-loop system, either in a deterministic sense for the $\tilde{A}_i$s, i = 1, N, separately, or in a stochastic sense for $[\tilde{A}_i , i = 1, N, \Pi]$. This should be analyzed in the manner of (Hopkins, 1987).

It can be observed that the state feedback regulator is obtained as a special case of theorem 3.5 when C_i, i = 1 to N, are invertible (x_t measured) : from (3.4.4.c) the optimal gains are the $\Gamma_i C_i = - Q_{2i}^{-1} B'_i \Lambda_i X_i C'_i C'^{-1}_i X_i^{-1} C_i^{-1} C_i = - Q_{2i}^{-1} B'_i \Lambda_i$; that is, precisely the gains of theorem 3.1 for which (3.4.4.a and c) uncouple. This is expected because output and feedback measurements are equivalent for C_i invertible.

Regarding algorithms to solve (3.4.4) or, better, (3.4.6) when it is valid, some possibilities are given in § 3.6 with examples.

It may happen that the static output regulator given by theorem 3.5 fails to produce the desired performance. After some simulations and tuning of the weighting matrices, the designer would then want to add dynamics to the feedback loop, that is, to look for an output driven dynamic compensator.

Knowing that the regulator has access to regime, a compensator with switching coefficients may be tried

$$u_t = \Gamma_i y_t + K_i z_t \quad \text{when } r_t = i$$

where $z_t \in \mathbf{R}^s$ is the state of the compensator

$$\dot{z}_t = F(r_t) z_t + G(r_t) y_t \qquad , z(t_0) = z_0$$

Using the approach of theorem 3.5 an optimization problem may be formulated to find the values of Γ_i, K_i, F_i and G_i, $i = 1$ to N, which produce the smallest quadratic cost. Averaging again allows us to transform this into a deterministic problem and the matrix maximum principle gives necessary conditions for optimality very similar to (3.4.4). If s is chosen equal to n, that is if the compensator has the same dimension as the plant itself, the optimization problem has an obvious connection with the Jump Linear Quadratic Gaussian (JLQG) situation which we shall study in chapter 5. The state of the compensator z_t can then be interpreted as the estimate of the state of the system delivered by an optimal full-order Luenberger observer, or, if some noise is present, by the corresponding Kalman filter. Even for $s < n$, the interpretation of the dynamic part of the compensator as a reduced order observer gives some insight into the structure of the optimal solution.

It may also happen that the constraints on the control law cannot be expressed by a single output relation like $y = Cx$. In a decentralized control situation, for example, the available state components vary from local station to local station since they are essentially formed from the subsystem state. A simple modification to the regulator of theorem 3.5 accommodates this situation.

For simplicity consider the case of m interconnected subsystems with scalar controls $u_{\ell t}$, $\ell = 1, m$. Define $z_\ell \in \mathbf{R}^{n_\ell}$ as the information available for the ℓ-th component of the control

$$z_{\ell t} = C_\ell(r_t)\, x_t \qquad \ell = 1, m$$

For decentralized control of interconnected subsystems C_ℓ would for example be the sparse matrix picking up the measured components of $x_{\ell t}$, the state of the l-th subsystem in x_t. The gains to be optimized are then the Γ_ℓs, $\ell = 1, m$, with

$$u_{\ell t} = \Gamma_\ell (r_t)\, z_{\ell t}$$

With Q_{2i} a diagonal matrix with non zero entries $q_{\ell i}$, $\ell = 1, m$ and $b_{\ell i}$ the ℓ-th column of B_i, it is readily verified that (3.4.4.c) specializes to

$$\Gamma_{\ell i} = -\,(1/q_{\ell i})\, b_{\ell i}'\, \Lambda_i\, X_i\, C_{\ell i}'\, (C_{\ell i}\, X_i\, C_{\ell i}')^{-1} \qquad , \ell = 1, m,\ i = 1, N$$

where we have used the notations of theorem 3.5.

3.5 The Optimal Non-Switching State Feedback

Let us now consider the converse of the situation in § 3.4: while the plant state x_t is available, the regime observation is no longer present. As in § 3.4, performance specifications may sometimes require the estimation of the missing variables in order to reduce uncertainty surrounding the operation of the plant, and chapters 5, 6 and 7 provide some results in this context. However, a simple constrained regulator is also interesting from the point of view of controller complexity, and the simplest regulator feeding back x_t for r_t unknown and not estimated would be the non-switching state feedback

$$u_t = \Gamma\, x_t \qquad \forall r_t \in S \tag{3.5.1}$$

With respect to the target tracking application of chapter 1, this would correspond to the actions of the tracker when it is unable to isolate the type and status of the target currently in its field of view, either because the algorithm has not converged yet or because the corresponding electronic circuits have failed. The question of the best choice for the single gain Γ in (3.5.1) can be formulated as an optimization problem

$$\underset{\Gamma}{\text{Min}}\ J \qquad \text{subject to } \dot{x}_t = A(r_t)\, x_t + B(r_t)\, u_t \tag{3.5.2}$$

$$\text{and to (3.5.1)}$$

which is solved with the same technique used for switching output feedback. Averaging is now over the initial state and regime distributions with $X_{i0}= E\{x_0 x_0' \phi_{t_0 i}\}$ and $P\{r_{t_0} = i\} = E\{\phi_{t_0 i}\}$. It is assumed that the initial regime distribution is not degenerate ($E\{\phi_{t_0 i}\} > 0$, i = 1 to N).

Theorem 3.6

A necessary condition for (3.5.1) to be optimal is that the gain Γ satisfies

$$\dot{X}_i = \tilde{A}_i X_i + X_i \tilde{A}'_i + \sum_{j=1}^{N} \pi_{ji} X_j \quad ; \; X_i(t_0) = X_{i0} \tag{3.5.3.a}$$

$$-\dot{\Lambda}_i = \tilde{A}'_i \Lambda_i + \Lambda_i \tilde{A}_i + Q_{1i} + \Gamma' Q_{2i} \Gamma + \sum_{j=1}^{N} \pi_{ij} \Lambda_j \quad ; \; \Lambda_i(t_f) = 0 \tag{3.5.3.b}$$

$$\sum_{i=1}^{N} Q_{2i} \Gamma X_i + \sum_{i=1}^{N} B'_i \Lambda_i X_i = 0 \tag{3.5.3.c}$$

where $\tilde{A}_i = A_i + B_i \Gamma$, i = 1, N.

Proof :

It proceeds as with theorem 3.5, except that there is now a single gain Γ as free parameter so that the third optimality condition becomes $\partial H/\partial \Gamma = 0$, where H is the hamiltonian of the problem. Using the gradients of appendix 2, this condition is computed to obtain (3.5.3.c). □

Equation (3.5.3.c) is linear in Γ and can be readily solved for known Λ_i and X_i. However, if the control weighting matrix is chosen as regime independent an even simpler explicit solution is

$$\Gamma = - Q_2^{-1} \left(\sum_{i=1}^{N} B'_i \Lambda_i X_i\right) \left(\sum_{i=1}^{N} X_i\right)^{-1}$$

The comments following theorem 3.5, in particular those concerning the asymptotic case, dynamic feedback and multiple constraints, have obviously direct counterparts in the present situation. Algorithms for solving the TPBVP (3.5.3) are the same as those needed needed for (3.4.4), and are given in § 3.6.

Finally, the results of § 3.4 and 3.5 may be combined to obtain the optimal non-switching output feedback

$$u_t = \Gamma\, y_t \qquad \forall r_t \in S$$

as the solution of

$$\dot{X}_i = \tilde{A}_i X_i + X_i \tilde{A}'_i + \sum_{j=1}^{N} \pi_{ji} X_j \qquad ; \; X_i(t_0) = X_{i0}$$

$$-\dot{\Lambda}_i = \tilde{A}'_i \Lambda_i + \Lambda_i \tilde{A}_i + Q_{1i} + C'_i \Gamma' Q_2 \Gamma C_i + \sum_{j=1}^{N} \pi_{ij} \Lambda_j \qquad ; \; \Lambda_i(t_f) = 0$$

$$\Gamma = -Q_2^{-1} \left(\sum_{i=1}^{N} B'_i \Lambda_i X_i C'_i\right) \left(\sum_{i=1}^{N} C_i X_i C'_i\right)^{-1}$$

for the case of a regime independent control weighting matrix.

3.6 Algorithms

The optimization and evaluation of regulators for JLQ systems studied in this chapter require the solution of various sets of matrix equations; coupled Riccati equations for the optimal JLQ regulator (3.2.5), coupled Liapunov equations for suboptimal solutions (3.3.3) or the TPBVPs (3.4.4) and (3.5.3) for the optimal feedback with constrained structures. The purpose here is to describe some of the algorithms available for solving these equations. This selection is undoubtedly biased and, moreover, no claim is made regarding numerical efficiency : the algorithms proposed could certainly be improved by incorporating more advanced numerical techniques, but this would be the subject of another volume.

Consider first the Riccati equations of the JLQ regulator, which are recalled for convenience

$$- \dot{\Lambda}_i = A_i' \Lambda_i + \Lambda_i A_i - \Lambda_i B_i Q_{2i}^{-1} B_i' \Lambda_i + Q_{1i} + \sum_{j=1}^{N} \pi_{ji} \Lambda_j \qquad (3.6.1)$$

$$\Lambda_i (t_f) = 0$$

A quasi-linearization and successive approximation algorithm was proposed in (Wonham, 1971). For some fixed $\Gamma_i^{(c)}$, i = 1, N, define $\Lambda_i^{(c+1)}$, i = 1, N, as the solutions of the set of coupled Liapunov equations

$$- \dot{\Lambda}_i^{(c+1)} = \tilde{A}_i^{(c)'} \Lambda_i^{(c+1)} + \Lambda_i^{(c+1)} \tilde{A}_i^{(c)} + \sum_{j=1}^{N} \pi_{ij} \Lambda_j^{(c+1)} + Q_{1i} + \Gamma_i^{(c)'} Q_{2i} \Gamma_i^{(c)}$$

$$\Lambda_i^{(c+1)} (t_f) = 0$$

with $\tilde{A}_i^{(c)'} = A_i + B_i \Gamma_i^{(c)}$, i = 1, N. Then modify $\Gamma_i^{(c)}$ into $\Gamma_i^{(c+1)} = - Q_{2i}^{-1} B_i' \Lambda_i^{(c+1)}$ and iterate the procedure. It can be shown that this algorithm produces a monotone decreasing sequence of positive semi-definite matrices

$$0 \leq \Lambda_i^{(c+1)} \leq \Lambda_i^{(c)} \qquad , i = 1, N$$

so that the limit as $c \to \infty$ exists. By the dominated convergence theorem, this limit is Λ_i, the desired solution of (3.6.1). An inconvenience of this algorithm is that no a priori estimate of the convergence rate is given, so that it is difficult to give a bound on the corresponding CPU time for a given precision.

As an alternative, we consider the use of an homotopy algorithm. The idea is simple (see the survey (Allgower and Georg, 1980) for an introduction to this technique): let us consider the problem, supposedly difficult, of finding a zero $\bar{\zeta}$ of a given function $F(\zeta)$. Instead of generating a sequence $\zeta^{(c+1)} = G(\zeta^{(c)})$ which would converge to $\bar{\zeta}$ as c goes to infinity, we choose a parametrization of F, $H(\zeta, \alpha)$, such that $H(\zeta,1) = F(\zeta)$ and that $H(\zeta,0) = 0$ has a trivial solution, say $\zeta(0)$. By following the continuation path $\alpha = 0 \to \alpha = 1$ the solution $\bar{\zeta}$ can then be obtained as $\zeta(1)$ computed by a numerical integration

$$\zeta(1) = \zeta(0) + \int_0^1 \left(\frac{d\zeta}{d\alpha}\right) d\alpha$$

where $d\zeta/d\alpha$ is found by differentiating the equation $H(\zeta, \alpha)= 0$

$$H_\zeta (\zeta(\alpha),\alpha) \frac{d\zeta}{d\alpha} + H_\alpha (\zeta(\alpha),\alpha) = 0$$

where H_ζ and H_α denote the partial derivatives. This equation must be solved for $d\zeta/d\alpha$ and special care must be exercised at points where $H_\zeta (\zeta(\alpha),\alpha)$ is singular. The ability to go from $\alpha = 0$ to $\alpha = 1$ along a connected path is a key condition for a simple implementation of homotopy algorithms, even though refinements have been proposed when the path is more complex. These algorithms are also sometimes called continuation, or embedding, algorithms.

If the coupling ($\sum \pi_{ij}\Lambda_j$) in (3.6.1) were to vanish then standard library routines could be used to find the Λ_is; since programs solving classical Riccati equations are now widely available. With this in mind , (3.6.1) is parameterized

$$-\dot{\Lambda}_i = A_i' \Lambda_i + \Lambda_i A_i - \Lambda_i B_i Q_{2i}^{-1} B_i' \Lambda_i + Q_{1i} + \alpha \sum_{j=1}^{N} \pi_{ij} \Lambda_j \tag{3.6.2}$$

$$\Lambda_i (t_f) = 0$$

At $\alpha = 0$ the above system becomes decoupled and the $\Lambda_i^{c=0}$ are easily computed through N runs of standard Riccati equation routines. We next introduce the matrices P_i , i = 1, N as $P_i = d\Lambda_i/d\alpha$, and find that the P_is satisfy a set of coupled Liapunov equations

$$\dot{P}_i = (A_i - B_i Q_{2i}^{-1} B_i' \Lambda_i)' P_i + P_i \ (A_i - B_i Q_{2i}^{-1} B_i' \Lambda_i) + \sum_{j=1}^{N} \pi_{ij} P_j + \sum_{j=1}^{N} \pi_{ij} \Lambda_j \tag{3.6.3}$$

$$P_i (t_f) = 0$$

Using an integration step-size $1/L (L \in \mathbf{N})$ the homotopy algorithm is as follows. Starting from the known $\Lambda_i^{c=0}$, i = 1, N, compute from (3.6.3) the corresponding $P_i^{c=0}$, and approximated the $\Lambda_i^{c=1}$, solutions of (3.6.1) for $\alpha = 1/L$ by

$$\Lambda_i^{c=1} = \Lambda_i^{c=0} + (1/L)\, P_i^{c=0} \qquad , i = 1, N$$

Iterating on c till c = L produces values for $\Lambda_i^{c=L}$, i = 1 to N, solutions of (3.6.2) for α = 1, that is, solutions of the original system.

There is obviously a compromise between the accuracy of the result obtained and the computation time: large values of L, that is a small step-size, improve the accuracy at the cost of an increased computation time (L iterations are needed to reach $\alpha = 1$). This algorithm is interesting because the number of iterations is fixed a priori.

Considering a simple example (n = 3, N = 3), a study of the precision and CPU time as a function of L was reported in (Mariton and Bertrand, 1985). The following figure summarizes the results.

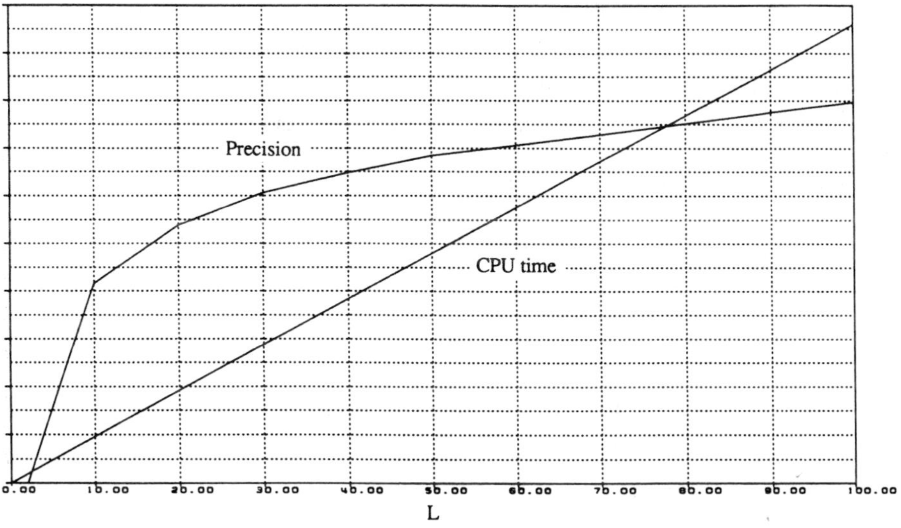

Figure 3.2 - Precision and CPU time/number of steps.

We see that for $L \geq 20$ the precision improvement becomes marginal with a linear growth of CPU time so that in this case a step size $1/L \sim 0.1$ would be advisable.

Other parameterizations of the original equation may also be selected. For example

$$- \dot{\Lambda}_i = A_i' \Lambda_i + \Lambda_i A_i - \Lambda_i B_i Q_{2i}^{-1} B_i' \Lambda_i + \alpha Q_{1i} + \frac{\alpha^2}{2} \sum_{j=1}^{N} \pi_{ij} \Lambda_j$$

This makes the initial step trivial, since for $\alpha = 0$ one readily computes $\Lambda_i^{c=0} = 0$, $i = 1$ to N.

Consider next coupled Liapunov equations of the form

$$- \dot{P}_i = A_i' P_i + P_i A_i + Q_{1i} + \sum_{j=1}^{N} \pi_{ij} P_j \tag{3.6.4}$$

$$P_i (t_f) = 0$$

This type of equation was encountered in chapter 2 when studying mean square stability, and in § 3.3 when the suboptimality degree of an arbitrary feedback law was computed : the value of the cost was deduced from a set like (3.6.4) with A_i replaced by $\tilde{A}_i = A_i + B_i \Gamma_i$ and Q_{1i} by $Q_{1i} + \Gamma_i' Q_{2i} \Gamma_i$. Note also that both the quasilinearization and homotopy algorithms for coupled Riccati equations use the solutions of a set of coupled Liapunov equations at very step . There are thus several reasons why it is interesting to learn how to integrate (3.6.4). In fact this is just a set of coupled linear differential equations in the entries of the P_is and many methods for solving them are available. One possibility is now explained.

As with the classical Liapunov equation, (3.6.4) can be written

$$- \dot{\underline{P}} = F \underline{P} + Q$$

$$\underline{P} (t_f) = 0 \tag{3.6.5}$$

using the following notations : for any matrix M, $\underline{M}$ denotes the column vector stacking the columns of M and $\underline{P} = [\underline{P}_1', \underline{P}_2', \ldots, \underline{P}_N']'$ and $Q = [\underline{Q}_{11}', \underline{Q}_{12}', \ldots, \underline{Q}_{1N}']'$. The $Nn^2 \times Nn^2$ matrix F is given by

$$F = \begin{bmatrix} F_1 & \pi_{12} I & \cdots & \pi_{1N} I \\ \pi_{21} I & F_2 & \cdots & \\ \cdot & & \cdot & \cdot \\ \cdot & & \cdot & \cdot \\ & & \cdot & \\ \cdot & & & \cdot \\ \pi_{N1} I & & \cdots & F_N \end{bmatrix}$$

with $F_i = I \otimes A_i' + A_i' \otimes I + \pi_{ii} I$, $i = 1, N$ where $\otimes$ designates the Kronecker product.

The solutions of (3.6.5) are readily obtained through

$$\underline{P}(t) = \exp Ft \left(\int_t^{t_f} \exp - Fs \, ds \right) \underline{Q}$$

and, if F is invertible

$$\underline{P}(t) = (-I + \exp F(t_f - t)) F^{-1} \underline{Q}$$

Finally consider the non linear matrix TPBVP to be solved for the optimal output feedback with switching gains

$$\dot{X}_i = \tilde{A}_i X_i + X_i \tilde{A}'_i + \sum_{j=1}^{N} \pi_{ji} X_j \quad ; \; X_i(t_0) = X_{i0} \tag{3.6.6.a}$$

$$-\dot{\Lambda}_i = \tilde{A}_i' \Lambda_i + \Lambda_i \tilde{A}_i + Q_{1i} + C_i' \Gamma_i Q_2 \Gamma_i' C_i + \sum_{j=1}^{N} \pi_{ij} \Lambda_j \quad ; \Lambda_i(t_f) = 0 \tag{3.6.6.b}$$

$$\Gamma_i = - Q_{2i}^{-1} B_i' \Lambda_i X_i C_i' (C_i X_i C_i')^{-i} \tag{3.6.6.c}$$

with $\tilde{A}_i = A_i + B_i \Gamma_i C_i$, $i = 1$ to N. Algorithms to solve (3.6.6) are easily adapted to the similar TPBVP for the optimal non-switching state output feedback.

A first possibility is to tackle the optimality conditions directly. At the initial step N matrices Λ_i^0 are chosen, e.g. those of the state feedback with switching gains. At step c (3.6.6.a) and (3.6.6.c) are solved as 2N coupled nonlinear equations in $X_i^{(c)}$ and $\Gamma_i^{(c)}$ for fixed $\Lambda_i^{(c)}$, i = 1 to N. The $\Lambda_i^{(c)}$s are then modified into the $\Lambda_i^{(c+1)}$s defined as the solutions of the N coupled linear equations (3.6.6.b) for fixed $X_i^{(c)}$ and $\Gamma_i^{(c)}$. Step c is iterated until convergence.

This is a rather complex algorithm and the computational burden of 2N nonlinear matrix equations at each step is not very encouraging. As an alternative, the following algorithm can be proposed. The initial step now consists of picking N gains Γ_i^0 such that $\tilde{A}_i + \frac{1}{2}\pi_{ii} I_n$ is stable (deterministically) for i = 1 to N. At step c (3.6.6.a) and (3.6.6.b) are solved as N decoupled linear equations in $X_i^{(c)}$ and $\Lambda_i^{(c)}$ for fixed $\Gamma_i^{(c)}$, i = 1 to N. The $\Gamma_i^{(c)}$s are modified into the $\Gamma_i^{(c+1)}$s defined by (3.6.6.c) for fixed $\Lambda_i^{(c)}$ and $X_i^{(c)}$. Step c is iterated until convergence.

This algorithm is much simpler than the previous one and can be used to solve problems of realistic dimensions (n, N). However we have to guess a constrained stabilizing gain at the initial step which is sometimes a difficult problem in itself. Until recently little was known about the convergence properties of the above two algorithms. Essentially the only result was a lemma from (Levine and Athans, 1970).

Lemma

For $\Gamma_i^{(c)}$, i = 1 to N, given by (3.6.6 c) for $X_i = X_i^{(c)}$, $\Lambda_i = \Lambda_i^{(c)}$ the solutions of (3.6.6.a) and (3.6.6.b) are noted $X_i^{(c+1)}$ and $\Lambda_i^{(c+1)}$. The following is then true

(i) If the matrices $\underline{\tilde{A}}_i^{(c)} = A_i - B_i \Gamma_i^{(c)} + \frac{1}{2}\pi_{ii} I_n$, i = 1 to N, are stable (in a deterministic sense) then there exists N unique positive definite matrices $\Lambda_i^{(c+1)}$.

(ii) If there exists N positive definite matrices $X_i^{(c)}$ then

$$\text{trace } \Lambda_i^{(c+1)} \leq \text{trace } \Lambda_i^{(c)} \qquad , i = 1 \text{ to } N$$

<u>Proof</u> :

The proof is essentially the same as that of (Levine and Athans, 1970) and is thus omitted. □

It can be observed that the algorithms and lemma recalled here are not very different from those of Levine and Athans for the deterministic LQ output feedback problem. The occurence of random regime transitions manifests itself solely through the coupling terms in the optimality conditions. The lemma does not prove convergence of the algorithms and at each step, a stability check must be performed.

A different approach is to consider the gain equation (3.6.6.c) as an equation for the gradient of the cost with respect to the desired gains. Indeed (3.6.6.c) is clearly the solution of $\partial J/\partial \Gamma_i = 0$, i = 1 to N, where

$$\frac{\partial J}{\partial \Gamma_i} = Q_{2i}\,\Gamma_i\,(C_i\,X_i\,C_i') - B_i'\,\Lambda_i\,X_i\,C_i' \qquad i = 1 \text{ to } N \tag{3.6.7}$$

Solving $\partial J/\partial \Gamma_i = 0$ couples the forward and backward integrations but an alternative is to compute the gradient and up-date the gains through a classical descent algorithm. Again, stabilizing gains must be chosen at the initial step. Then, at step c, (3.6.6.a) and (3.6.6.b) are solved as N coupled linear equations in $X_i^{(c)}$ and $\Lambda_i^{(c)}$ for fixed $\Gamma_i^{(c)}$, i = 1 to N, respectively. We then evaluate the gradient $(\partial J/\partial \Gamma_i)^{(c)}$ according to (3.6.7) and modify the $\Gamma_i^{(c)}$s into the $\Gamma_i^{(c+1)}$s as

$$\Gamma_i^{(c+1)} = \Gamma_i^{(c)} - \alpha\,(\partial J/\partial \Gamma_i)^{(c)} \qquad i = 1 \text{ to } N$$

where α is the gradient step. This is iterated until convergence.

The advantage of this algorithm is its simplicity. The need to guess an initial stabilizing gain remains but the freedom to choose α makes it easier to satisfy stability at each step. As any descent algorithm it suffers the risk of stopping at a local minimum so that it is certainly advisable to perform several runs with different initializations.

Reference to more recent results on constrained otptimization are given in § 3.7. A survey is (Mäkkila and Toivonen, 1987) which is pertinent here even though its original scope is limited to deterministic models.

To illustrate constrained designs, two simple numerical examples are now presented.

Example 1 :

Consider a system in $\mathbf{R}^2$ with two regimes and $\Pi = \begin{bmatrix} 0. & 0. \\ 1. & -1. \end{bmatrix}$. The observation is scalar and matrices are given as

$$A_1 = \begin{bmatrix} -1. & 0.5 \\ 0. & 1. \end{bmatrix} \quad B_1 = \begin{bmatrix} 0. \\ 1. \end{bmatrix} \quad C_1 = [0.\ 1.]$$

$$A_2 = \begin{bmatrix} 0.25 & 0.5 \\ 0. & 0.25 \end{bmatrix} \quad B_2 = \begin{bmatrix} 0. \\ 1. \end{bmatrix} \quad C_2 = [0.\ 1.]$$

The cost index is computed for unitary weighting coefficients and only the asymptotic case is discussed. The optimal state feedback with switching gains is computed first :

$$\Lambda_1 = \begin{bmatrix} 0.4949 & 0.1010 \\ - & 2.4495 \end{bmatrix} \text{ and } \Lambda_2 = \begin{bmatrix} 2.3548 & 0.5635 \\ - & 1.7688 \end{bmatrix}$$

gives state feedback gains

$$\Gamma_1 = [0.1010\ \ 2.4495] \text{ and } \Gamma_2 = [0.5635\ \ 1.7688]$$

Using the second algorithm the output feedback solution with switching gains is computed next. As c becomes large convergence occurs to co-state matrices

$$\Lambda_1^\infty = \begin{bmatrix} 0.5 & 0.0995 \\ - & 2.4503 \end{bmatrix} \text{ and } \Lambda_2^\infty = \begin{bmatrix} 3.0 & 0.6577 \\ - & 1.7967 \end{bmatrix}$$

with (scalar) output feedback gains

$$\Gamma_1^\infty = 2.5135 \qquad \text{and } \Gamma_2^\infty = 1.9319$$

For $X_{10} = \bar{p}_{10} I_2$ and $X_{20} = \bar{p}_{20} I_2$ where $\bar{p}_{10} = 1 - \bar{p}_{20} = \mathcal{P}\{r_{t_0} = 1\}$ the optimal state and output feedback costs are given by

$$J^* = \bar{p}_{10} \operatorname{trace} \Lambda_1 + (1 - \bar{p}_{10}) \operatorname{trace} \Lambda_2$$
$$J = \bar{p}_{10} \operatorname{trace} \Lambda_1^\infty + (1 - \bar{p}_{10}) \operatorname{trace} \Lambda_2^\infty$$

so that the suboptimaly introduced by the output constraints is summarized by the following table where $\upsilon = \dfrac{J - J^*}{J^*}$

The worst case (E $\{\phi_{01}\}$ = 1.) leads to a performance degradation smaller than 20% which might often be judged good enough to accept a static output feedback solution without estimating the missing state variable.

Example 2 :

Consider a system in $\mathbf{R}^2$ with two regimes and transition rates and cost weightings as in example 1. Matrices are given as

$$A_1 = \begin{bmatrix} -1. & 0.75 \\ 0.5 & 1. \end{bmatrix} \qquad B_1 = \begin{bmatrix} 0. \\ 1. \end{bmatrix} \qquad C_1 = [0.\ 1.]$$

$$A_2 = \begin{bmatrix} 0.5 & 0.25 \\ 0.5 & 0.25 \end{bmatrix} \qquad B_2 = \begin{bmatrix} 0. \\ 1. \end{bmatrix} \qquad C_2 = [0.\ 1.]$$

The reference state feedback solution in the asymptotic case ($t_f \to \infty$) is characterized by

Table 3.1 - Degree of suboptimality.

p_{01}	1.	.5	0.
J^S	2.9444	3.5340	4.1236
J^0	2.9503	3.8735	4.7967
υ(%)	.2	9.6	16.3

$$\Lambda_1 = \begin{bmatrix} 0.6109 & 0.6677 \\ - & 2.7325 \end{bmatrix} \quad \text{and} \quad \Lambda_2 = \begin{bmatrix} 0.9263 & 0.5907 \\ - & 1.7725 \end{bmatrix}$$

$$\Gamma_1 = [\,0.6677 \quad 2.7325\,] \quad \text{and} \quad \Gamma_2 = [\,0.5907 \ 1.7725\,]$$

The gradient algorithm is then used to compute the (scalar) gains of regime dependent output feedback. Figure 3.3 plots the gains $\Gamma_1^{(c)}$ along iterations for several initial guesses. For this example there is a single minimum in a rather large vicinity of the gains obtained.

$$\Gamma_1^{\infty} = 3.5755 \qquad \text{and} \quad \Gamma_2^{\infty} = 2.0990$$

The suboptimality degree, computed as for example 1, also remains around 20% in the worst case.

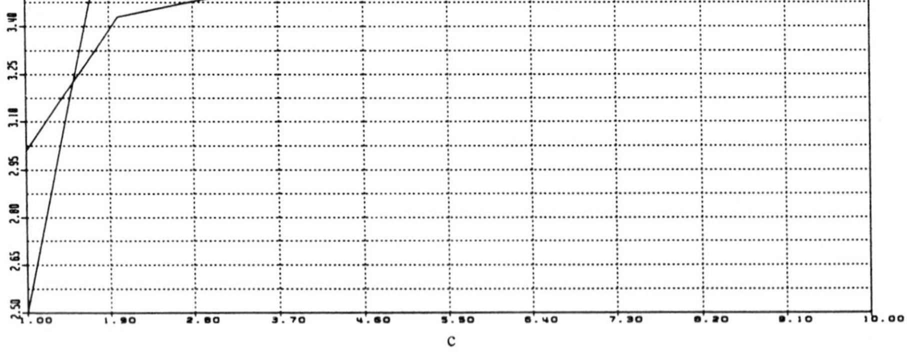

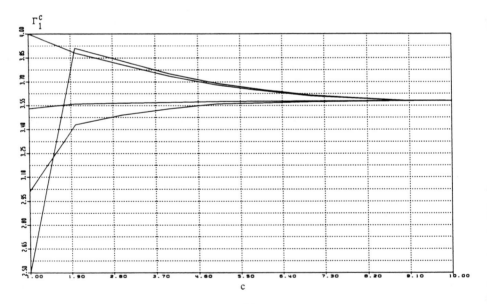

Figure 3.3 - Convergence of the gradient algorithm.

3.7 Notes and References

After early results of Florentin, Kats and Krasovskii, the papers of Sworder (1969) and Wonham (1971) marked the beginning of a renewed interest in the control problem for hybrid models. The many contributions of Sworder, extending his 1969 paper in various directions, are illustrated in forthcoming chapters.

The Jump Linear Quadratic control problem was considered here without a penalization on the terminal error x_{t_f}. In some situations, like the target tracking application in chapter 1, the terminal error must be given special emphasis and this can be achieved by modifying the cost (3.2.1) into

$$x_{t_f}' Q_3 x_{t_f} + \int_{t_0}^{t_f} (x_t' Q_1 x_t + u_t' Q_2 u_t) dt$$

and averaging over x_{t_0} and r_{t_0}. The terminal weighting matrix may be chosen as regime dependent $(Q_3 = Q_3(r_t))$ if needed.

One difficulty with the constrained design of § 3.4 and 3.5 concerns stabilization. It is not easy to establish conditions under which the asymptotic optimal constrained regulator guarantees some kind of stability. This is in contrast to the full state and regime feedback situation where two such results were obtained in §3.2.3 leading to deterministic and stochastic stabilization. It seems however that the problem can be better understood along the lines of (Hopkins, 1987) and (Mäkkila and Toivonen, 1987) but this is not pursued here.

There is a strong similarity between the constrained JLQ regulators studied here and the deterministic constrained LQ regulators found in the literature. A key reference is (Levine and Athans, 1970). The explicit integration of the coupled Liapunov equations (3.6.5) was originally presented as (Jodar and Mariton, 1987). Further discussions of the numerical and stabilization aspects are (Kosut, 1970, Anderson and Moore, 1971, Kwakernaak and Sivan, 1972, Srinivasa and Rajagopalou, 1979). Additional results and references may be found in the detailed survey (Mäkkila and Toivonen, 1987).

An alternative approach to constrained regulator design is to preserve some eigensubspace of the optimal closed-loop dynamics under unconstrained feedback. This was pioneered in (Hopkins et al., 1981, Medanic and Uskokovic, 1983, Medanic et al., 1985) for the LQ problem and could easily be extended to the JLQ setting. Similarly the optimal projection approach by (Bernstein and Hyland, 1985 a and b) provides means to arrive at constrained JLQ regulators.

The simple examples of § 3.6 demonstrate that the loss of optimality due to constraining the regulator might be quite tolerable. It is then not necessary to estimate the missing variables to meet performance requirements.

The next step is to study how much robustness can be achieved with the JLQ design and, more fundamentally, to understand better what the notion of robustness might be for a system with stochastic jumps. This is the purpose of chapter 4.

References

Allgower, E., and Georg, K. (1980). Simplicial and continuation methods for approximating fixed points and solutions to systems of equations, SIAM Review, 22 : 28.

Anderson, B.D.O., and Moore, J.B. (1971). Linear Optimal Control, Prentice-Hall, New-York.

Bernstein, D.S., and Hyland, D.C. (1985a). The optimal projection equations for reduced order state estimation, IEEE Trans. Aut. Control, AC-30 : 583.

Bernstein, D.S., and Hyland, D.C. (1985b). The optimal projection/maximum entropy approach to design low-order, robust controller for flexible structures, Proc. 24th IEEE Conf. Decision Control, Fort-Lauderdale, pp. 745-752.

Hopkins, W.E.Jr. (1987). Optimal stabilization of families of linear stochastic differential equations with jump coefficients, SIAM J. Control Optim., 25 : 1587.

Hopkins, W.E.Jr., Medanic, J., and Perkins, W.R. (1981). Output feedback pole placement in the design of suboptimal linear quadratic regulators, Int. J. Control , 34 : 593.

Ji, Y. and Chizeck, H.J. (1987). Controllability, observability and the jump linear quadratic problem in continous-time, Proc. 26th IEEE Conf. Decision Control, Los Angeles, pp. 329-331.

Jodar, L., and Mariton, M. (1987). Explicit solutions for a system of coupled Liapunov differential matrix equations, Proc. Edinburgh Math. Society, series II, 30 : 427.

Kosut, R.L. (1970). Suboptimal control of linear time invariant systems subject to control structure constraints, IEEE Trans. Aut. Control, AC-15 : 557.

Kwakernaak, H., and Sivan, R. (1972). Linear Control Systems, J. Wiley, New-York.

Levine, W.S., and Athans, M. (1970). On the determination of the optimal output feedback gains for linear multivariable systems, IEEE Trans. Aut. Control, AC-15 : 44.

Mäkilä, P.M., and Toivonen, H.T. (1987). Computational methods for parametric LQ problems - A survey, IEEE Trans. Aut. Control, AC-32 : 658.

Mariton, M., and Bertrand, P. (1985). A homotopy algorithm for solving coupled Riccati equations, Opt. Control Appl. Methods, 6 : 351.

Medanic, J., and Uskokovic, Z. (1983). The design of optimal output regulators for linear multivariable systems with constant disturbances, Int. J. Control, 37 : 809.

Medanic, J., Petranovic, D. and Gluhajic, N. (1985). The design of output regulators for discrete-time linear systems by projective control, Int. J. Control, 41 :615.

Rudin, W. (1966). Real and Complex Analysis, Mc Graw-Hill, New-York.

Srinivasa, Y.G., and Rajagopalou, T. (1979). Algorithms for the computation of optimal output feedback gains, Proc. 18th IEEE Conf. Decision Control, Fort-Lauderdale, pp. 576-579.

Sworder, D.D. (1969) Feedback control for a class of linear systems with jump parameters, IEEE Trans. Aut. Control, AC-14 : 9.

Wonham, W.M. (1971). Random differential equations in control theory, Probabilistic Methods in Applied Mathematics (A.T. Bharucha-Reid, ed.), vol. 2, Academic Press, New-York, p. 131.

4

Robustness

4.1 Introduction

When dealing with engineering problems, we should keep in mind the approximate nature of the modeling process: the choice of a representation is usually governed by mathematical tractability as much as by accuracy, and the parameters of this representation can in turn be estimated from the available plant data with only limited precision. For the solar thermal receiver example in chapter 1, the finite dimensional model is constituted from the patching of elementary lumped models, thus approximating the infinite dimensional laws of thermodynamics which describe the process more precisely. The movement of clouds over the heliostats is assumed markovian for the sake of simplicity, and the value of the transition rates must be associated with confidence intervals given by the numerical resolution that can be achieved from the available meteorological data.

The various studies aimed at assessing the influence of modeling inaccuracies on the controlled plant are grouped under the broad heading of robustness. As is often the case with engineering concepts, a precise definition of robustness is not easily formalized and we have to settle for vague assertions like "a design is said to be robust if large discrepancies between the model used and reality can be tolerated without unacceptable performance degradation."

It is then apparent that robustness, ill-defined as it may be, is a key asset for a theory that claims to be useful in the uncertain world of applications: a non-robust optimal control law may produce catastrophic results when applied to a real plant that differs from the design model in such a way that the ideally optimal solution destabilizes the true plant.

However, a brief analysis makes it clear that uncertainty in a stochastic system has two separate origins. The first is the modeling-related uncertainty discussed above:

the parameters of the model usually appear in the form of nominal values plus confidence bounds and the markovian linear finite dimensional form approximates reality which in general is non-markovian, non-linear and infinite dimensional. The corresponding aspects of robustness are well documented in the literature (see for example (Safonov 1980)).

The second kind of uncertainty, which is less often considered but which constitutes an unavoidable issue in stochastic control, is the robustness with respect to the realizations of the underlying stochastic process. The control solutions given in chapter 3 were conveniently designed to minimize the average value of the cost function but, once applied to the real plant, we observe a realization or sample path performance and not an averaged one. In other words, the performance of the system on a sample path, in which we are ultimately interested, is a random variable and the minimization of the mean of this variable guarantees a priori nothing about its variance or its distribution. Typically, in an economic planning problem, the policy maker does try to maximize the expected return of his policy between consecutive elections, but he also strives to maintain the instantaneous shape of the economy within acceptable limits to avoid bankruptcy before the end of his term (including!).

As another example, let us consider the situation of a starving man to whom the following alternative is proposed: either he receives $100 right away or he tosses a coin and receives $200 if it falls heads and nothing if it falls tails. Regarding only the mean reward the two terms of the alternative are equivalent, but if the man is really starving he will take the safer option and choose not to bet.

To design a regulator for a hybrid system a natural cost function is

$$J = \int_{t_0}^{t_f} (x'Q_1x + u'Q_2u)\, dt \qquad (4.1.1)$$

but this is a random variable and in chapter 3 we decided to replace it by the averaged cost

$$J = E\{ \int_{t_0}^{t_f} (x'Q_1x + u'Q_2u)\, dt \mid x_0, i_0, t_0 \} \qquad (4.1.2)$$

The questions studied in this chapter are thus: how does the minimization of the mean (4.1.2) affect the distribution of (4.1.1)? What new control strategies may be defined to influence the distribution directly? What is the variance of (4.1.1) under a control law optimal in the sense of (4.1.2)?

4.2 Variance of the Realized Cost

If a designer is not reasonably content with such a limited measure of performance as an average cost, other parameters must be provided from the realized cost distribution in order to allow a rational decision on the acceptability of a control policy. The first parameter we may think of is the cost variance, indicating to what extent performance is spread around its mean value. The variance carries significant information, as illustrated by the following manufacturing quality control problem.

If the manufacturer has contracted to deliver products above a certain minimum quality level, a key profit factor is the ability to obtain a sharp production quality distribution curve where the mean quality can be pushed close to the rejection threshold thanks to a small variance. If a better mean quality comes with an increased variance it is of no interest, and, in order to avoid a high rejection rate, most products must be produced significantly above the contracted quality level, thus inducing reduced benefits.

It can be expected, however, that higher order characteristics of the performance distribution are more difficult to compute than the mean. To simplify the problem, we will now restrict ourselves to a sub-class of hybrid models.

It is assumed that the regime markov chain is directed, that is, that the matrix of transition rates Π is band diagonal with $\pi_{i-1\,i} = -\pi_{ii} > 0$, $\pi_{11} = 0$. All the regimes are then transient except $r_t = 1$ which is absorbing and the chain jump destinations are deterministic ($i \rightarrow i\text{-}1 \rightarrow i\text{-}2$ ... , hence the term "directed" chain) with random jump instants $t_{i\,i-1}$, $t_{i-1\,i-2}$, ... Though restrictive, this assumption is quite realistic for some applications. In the fault-tolerant control problem for example, it corresponds to the absence of maintenance when the system, initially in the normal regime $r_t = N$, jumps through N, N-1, ... as partial failures occur until it reaches the complete failure regime, $r_t = 1$, where it is trapped.

With this assumption the variance of the random cost can be computed under a regime-dependent (switching) state feedback

$$u_t = \Gamma_i x_t \qquad \text{when } r_t = i .$$

Denoting by X_t and R_t the σ-algebras $\sigma - \{x_s, t_0 \le s \le t\}$ and $\sigma - \{r_s, t_0 \le s \le t\}$, the "instantaneous cost" is defined as the conditional expectation

$$J_t = E\,\{J \mid X_t \vee R_t\} \tag{4.2.1}$$

The mean cost optimized in chapter 3 is J_{t_0} and the realized cost J_{t_f}. The instantaneous cost therefore captures the performance of the process as it is perceived by the controller at time t.

For the present, the evolution of J_t is related to the variance of the cost $\text{var}J = E\{(J_{t_f} - J_{t_0})^2 \mid X_{t_0} \vee R_{t_0})\}$. First note that J_t is by definition adapted to $X_t \vee R_t$ and that, for $t \ge s$,

$$\begin{aligned} E\,\{J_t \mid X_s \vee R_s\} &= E\,\{E\,\{J \mid X_t \vee R_t\} \mid X_s \vee R_s\} \\ &= E\,\{J \mid X_s \vee R_s\} \\ &= J_s \end{aligned}$$

so that J_t is an $X_t \vee R_t$ - martingale. At time t, this martingale can be decomposed into two terms, $J_t = J_{at} + J_{pt}$, where J_{at}, an increasing adapted process, reflects the cost due to past actions, known at time t, and J_{pt}, called a potential from Meyer's decomposition theorem (Garsia, 1973), represents the expectation of future costs. From the computation of the cost given in chapter 3 (theorem 3.4) it is seen that

$$J_t = \int_{t_0}^{t} (x'Q_1x + u'Q_2u)\, d\tau \; + x_t'P_i\, x_t \qquad \text{when } r_t = i \tag{4.2.2}$$

where the integral is identified to J_{at} and the second term to J_{pt}. We recall from chapter 3 that the P_is, i = 1 to N, satisfy a set of coupled Liapunov equations

$$-\dot{P}_i = \tilde{A}_i'P_i + P_i\tilde{A}_i + Q_{1i} + \Gamma_i'\, Q_{2i}\,\Gamma_i + \sum_{j=1}^{N} \pi_{ij}\, P_j$$

$$P_i\,(t_f) = 0 \tag{4.2.3}$$

with $\tilde{A}_i = A_i + B_i\Gamma_i$, i = 1 to N. The integral for J_{at} is continuous while the potential J_{pt} has discontinuities over $[t_0, t_f]$. At points where J_t is continuous $dJ_t \triangleq J_{t+dt} - J_t$ is of order dt and therefore

$$\lim_{dt \to 0} dJ_t^2 = dJ_{pt}^2$$

Consider the instant, noted $t_{i+1\,i}$ where the regime jumps from i+1 to i. From (4.2.2) we see that

$$dJ_{pt_{i+1\,i}} = x'_{t_{i+1\,i}}\,(\,P_i\,(t_{i+1\,i}) - P_{i+1}\,(t_{i+1\,i}))\; x_{\,t_{i+1\,i}}$$

because it is assumed here that the state trajectory is continuous (see chapters 5 and 6 for a different situation). Adding the contributions over $[t_0, t_f]$ we find that

$$\text{var } J = E\{\sum_{t\in[t_0,t_f]} dJ_t^2 \mid X_{t_0} \vee R_{t_0}\}$$

$$= E\{\sum_{i=1}^{i_0-1} dJ_{pt_{i+1\,i}}^{\;2} \mid X_{t_0} \vee R_{t_0}\}$$

where $r(t_0) = i_0$. It is now convenient to assume that $t_f - t_0$ is large enough and that the coupled Liapunov equations for the P_is have reached their stationary values (the gains Γ_i, i = 1 to N, clearly have to be stabilizing in the mean square sense, see the discussion in the previous chapter on the asymptotic JLQ regulator for conditions justifying this assumption). The matrices P_i, i = 1 to N, can then be considered as constant with time and the (random) transition matrix of the system $\varphi(t)$ $(x_t = \varphi(t)\ x_0$ when $x_{t_0} = x_0)$ is obtained as

$$\varphi(t) = \varphi_i\,(t - t_{i+1\,i})\;\varphi_{i+1}\,(t_{i+1}) \ldots \varphi_{i_0}(t_{i_0}) \qquad t_{i+1\,i} \leq t < t_{i\,i-1}$$

where $t_i = t_{i\,i-1} - t_{i+1\,i}$ is the sojourn time in regime i and $\varphi_i(\tau) = \exp \tilde{A}_i\tau$. With these preliminaries, the variance expression follows.

Theorem 4.1

Under the assumption of a directed Markov chain and asymptotic performance evaluation $(t_f \rightarrow \infty,\ t_0 = 0)$ the cost variance is

$$\text{var } J = \sum_{i=1}^{i_0} \prod_{j=1}^{i_0} |\pi_{jj}| \int_0^\infty \cdots \int_0^\infty (x_0'\bar{\varphi}_{i_0}(\tau_{i_0})' \ldots \bar{\varphi}_i(\tau_i)'\Delta P_{i-1}\bar{\varphi}_i(\tau_i) \ldots (\bar{\varphi}_{i_0}(\tau_{i_0})x_0)^2\, d\tau_{i_0} \ldots d\tau_i$$

where $\Delta P_i = P_{i+1} - P_i$ and $\bar{\varphi}_i(\tau) = \varphi_i(\tau) \exp \pi_{ii}\,\tau/4$, i = 1 to N.

Proof:

Start with the simplest case when $i_0 = 2$. A single jump is then possible (from r = 2 to r = 1 at time t_{21}) and then

$$\text{var } J = E\{dJ^2_{pt_{21}} \mid X_{t_0} \text{ v } R_{t_0}\}$$

with

$$dJ_{pt_{21}} = x_0'\varphi_2(t_2)'(P_1 - P_2)\varphi_2(t_2)x_0$$

But t_2 is exponentially distributed with constant π_{22} so that, for $t_0 = 0$ and t_f large, the expectation can be evaluated explicitly

$$\text{var } J = |\pi_{22}| \int_0^\infty (\exp \pi_{22}\tau)(x_0'\varphi_2(\tau)'(P_1 - P_2)\varphi_2(\tau)x_0)^2\, d\tau$$

or, with the notations of the theorem

$$\text{var } J = |\pi_{22}| \int_0^\infty (x_0'\bar{\varphi}_2(\tau)\ \Delta\ P_1\bar{\varphi}_2(\tau)\ x_0)^2\, d\tau.$$

Next take $i_0 = 3$. There are the two contributions to the variance, when the regime jumps from 3 to 2 and 2 to 1 at t_{32} and t_{21} respectively. At time t_{32}

$$dJ_{pt_{23}} = x_0'\varphi_3(t_3)'(P_2 - P_3)\varphi_3(t_3)x_0$$

while at time t_{21}

$$dJ_{pt_{21}} = x_0'\varphi_3(t_3)'\varphi_2(t_2)'(P_1 - P_2)\varphi_2(t_2)\varphi_3(t_3)x_0$$

As the sojourn times t_2 and t_3 are independent random variables with exponential distribution the expectation is evaluated as

$$\text{var } J = |\pi_{33}| \int_0^\infty (x_0'\varphi_3(\tau)\,\Delta\, P_2\varphi_3(\tau)\, x_0)^2\, d\tau$$

$$+ |\pi_{33}|\,|\pi_{22}| \int_0^\infty \int_0^\infty (x_0'\varphi_3(\tau)'\varphi_2(\tau')'\Delta P_1\varphi_2(\tau')\varphi_3(\tau)\, x_0)^2\, d\tau\, d\tau'$$

which proves the theorem for $i_0 = 3$. Proceeding by induction on i_0, the general case is obtained in a similar manner. □

The theorem was proved under an arbitrary (stabilizing) control law, but it is immediately specialized if the optimal JLQ regulator is used, the only modification being to replace ΔP_i by $\Delta\Lambda_i = \Lambda_{i+1} - \Lambda_i$ where the Λ_i, $i = 1$ to N, are the solutions of the corresponding set of coupled Riccati equations (see theorem 3.1).

The variance equation is rather involved and, for i_0 large, it quickly becomes very heavy to compute. However, it is in a form suitable for computer implementation with integrals of exponential functions which can be mechanized to advantage using a formal manipulation language. It is also interesting to solve some simple situations that might give an insight into the general case, as we can see with the following example.

Example:

Consider a system in $\mathbf{R}^2$ with two regimes, $\Pi = \begin{bmatrix} 0. & 0. \\ 2. & -2. \end{bmatrix}$ and

$$A_1 = \begin{bmatrix} 2 & 1 \\ 0 & 2 \end{bmatrix} \quad B_1 = I_2 \qquad A_2 = \begin{bmatrix} -2.25 & 0 \\ 2.25 & 0.75 \end{bmatrix} \quad B_2 = I_2$$

with cost weighting matrices $Q_{11} = Q_{12} = I_2$, Q_{21} = diag (2,3), $Q_{22} = 2I_2$ and initial condition $x_0 = (0\ 1)'$. The mean optimal solution (i.e. the JLQ regulator given in chapter 3) is given by

$$\Lambda_2 = \begin{bmatrix} 3. & 1. \\ - & 4. \end{bmatrix} \qquad \Lambda_1 = \begin{bmatrix} 1. & 1. \\ - & 2. \end{bmatrix}$$

and the minimal mean is

$$E = \{J \mid i_0 = 2\} = 4$$

Using theorem 4.1, we see that the variance is then

$$\text{var } J = 0.8$$

It is interesting to compare this mean optimal feedback to a non-switching regulator using a gain $-Q_{21}^{-1}B_1'\Lambda_1$ whatever the value of r_t

$$u_t = -Q_{21}^{-1}B_1'\Lambda_1 x_t \quad \forall r_t = 1 \text{ or } 2$$

Using the Liapunov equations (3.3.3), the mean cost is found to be

$$E\{J \mid i_0 = 2\} \approx 4.2008$$

and the variance (from Theorem 4.1)

$$\text{var } J \approx 0.2552.$$

Not surprisingly the mean is degraded by using the non-optimal gain and the suboptimality degree of chapter 3 measures a degradation $\upsilon = 5\%$. However, it is also apparent that the variance is significantly smaller when using the arbitrary gain and in practice the designer may choose to tolerate a small degradation in the mean performance (5%) for the benefit of a strong reduction (70%) in the variance.

This simple example demonstrates that the mean optimal cost can indeed lead to a degradation of the variance. A knowledge of the entire cost distribution would be needed to understand this point more completely, and this is given in §4.4. However, before turning to this, we shall examine the notion of instantaneous cost introduced in (4.2.1) in more detail.

4.3 Bounds on the Instantaneous Cost

At time $t = t_0$ the designer ignores the sample path of the regime stochastic jump process that will govern the process evolution over $[t_0, t_f]$. It therefore seems natural to replace the cost (4.1.1) by its expectation (4.1.2) using an a priori model for the regime transitions. However, it was shown in §4.2 that exclusive concern with this mean performance should not serve as a single evaluation measure, because it can mask the unsatisfactory behavior of other performance parameters, e.g. the cost variance. The instantaneous cost $J_t = E\{J \mid X_t \vee R_t\}$ provides another valuable evaluation measure: even if it is chosen to operate with a mean optimal regulator (minimizing J_{t_0}), the computation of J_t as more information becomes available can indicate a deviation from "mean" operating conditions and the occurrence of adverse environmental disturbances. It may then be decided to revert to a safety back-up solution or even, for applications where this is possible, to abort process operation.

For jump linear systems under switching state feedback, we demonstrate the computation of bounds on the instantaneous cost which might reinforce the designer's confidence in a mean-based optimization, or, conversely, direct his attention to the possibility of poor sample path behavior.

Based on the property of J_t observed in §4.2, martingale inequalities (Garsia, 1973) provide the adequate tools to bound the instantaneous cost value. However, the computation of these bounds should remain as simple as possible and we also wish to express them in terms of a few parameters like the mean and variance.

Attention is once again restricted to systems with a directed chain in the asymptotic case so that theorem 4.1 can be used. Applying the notations in §4.1, the maximal cost is defined as

$$J_M = \sup_{t \in [t_0, t_f]} J_t$$

This measures the deterioration in performance as the process evolves on the control interval $[t_0, t_f]$. The knowledge of J_M may reassure the designer before a mean-based regulator is applied by showing that the probability of a severely degraded performance is quite low.

The next result is then useful for deriving a simple upper-bound on the probability that the maximal cost exceeds any given threshold.

Theorem 4.2

For λ a positive constant, the instantaneous cost is bounded by

$$\mathscr{P}\{J_M \geq \lambda\} \leq \text{Min}\ \{ \frac{1}{\lambda}\ x_0'P_{i_0}x_0,\ \frac{20\ \|v\|_2^2 + 2\sqrt{20}\ \|v\|_2\ x_0'P_{i_0}x_0}{(\lambda - \sqrt{10}\ \|v\|_1 - x_0'P_{i_0}x_0)^2}\ \} \quad (4.3.1)$$

where $r_{t_0} = i_0$, $v^2 = \sum_{i=1}^{i_0-1} dJ_{pt_{i+1}\ i}^{\ 2}$ and $\|.\|_p = E\ \{(.)^p\}^{1/p}$.

Proof:

To relate J_M to the mean and variance introduce

$$\tilde{J}_M = \sup_{t \in [t_0, t_f]} |J_t - J_{t_0}|$$

It is then clear that

$$J_M \leq \tilde{J}_M + J_{t_0}$$

and it follows from Minkowski's inequality that

$$\|J_M\|_p \leq \|\tilde{J}_M\|_p + J_{t_0}$$

Another martingale inequality (Garsia, 1973) is used to obtain the next bound

$$\|J_M\|_p \leq \sqrt{10p}\ \|v\|_p + J_{t_0} \quad \text{for } p = 1,2$$

For p = 1 this gives

$$E\{J_M\} \leq \sqrt{10}\ \|v\|_1 + J_{t_0}$$

and for p = 2

$$\operatorname{var} J_M \leq 20\ \|v\|_2^2 + J_{t_0}^2 + 2\sqrt{20}\ \|v\|_2 - E\{J_M\}^2 \leq 20\ \|v\|_2^2 + 2\sqrt{20}\ \|v\|_2.$$

Now, by Chebyschev's inequality,

$$\mathcal{P}\{J_M \geq \varepsilon + E\{J_M\}\} \leq \frac{\operatorname{var} J_M}{\varepsilon^2}$$

which can be modified into

$$\mathcal{P}\{J_M \geq \varepsilon + \sqrt{10}\ \|v\|_1 + J_{t_0}\} \leq \frac{1}{\varepsilon^2}\ (20\ \|v\|_2^2 + 2\sqrt{20}\ \|v\|_2 J_{t_0}).$$

With $J_{t_0} = x_0'P_{i_0}x_0$, $\varepsilon = \lambda - \sqrt{10}\ \|v\|_1 - J_{t_0}$, and rearranging terms, this gives the second bound in (4.3.1). The first bound is deduced from the classical super martingale inequality

$$\mathcal{P}\{J_M \geq \lambda\} \leq \frac{1}{\lambda}\ J_{t_0}$$

which can here be written as

$$\mathcal{P}\{J_M \geq \lambda\} \leq \frac{1}{\lambda}\ x_0'P_{i_0}x_0$$

completing the proof. □

To compute the bound given by (4.3.1) the norms $\|v\|_p$ must be evaluated for p = 1 and p = 2. With the definition of v this evaluation is directly related to the variance computation of §4.2 and, generalizing theorem 4.1 for $p \in \mathbf{N}$, we have

$$\|v\|_p^p = \sum_{i=1}^{i_0} \prod_{j=1}^{i_0} |\pi_{jj}| \int_0^\infty \int_0^\infty |x_0'\varphi_{i_0p}(\tau_{i_0})' \ldots \varphi_{ip}(\tau_i)'\Delta P_{i-1}\varphi_{ip}(\tau_i)(\varphi_{i_0p}(\tau_{i_0})x_0|^p\, d\tau_{i_0} \ldots\, d\tau_i$$

where $\Delta P_i = P_{i+1} - P_i$ and $\varphi_{ip}(\tau) = \varphi_i(\tau) \exp \pi_{ii}\tau/2p$, i = 1 to N. The norm $\|.\|_p$ is clearly awkward to compute and for this reason theorem 4.2 uses only its value for p = 1 an 2. It is true, however, that using higher order moments of v would produce tighter bounds. The merit of theorem 4.2 is thus in providing a simple bound involving only J_{t_0}, $\|v\|_1$ and $\|v\|_2^2$ (this last term was already given in §4.2), which gives a preliminary idea of the distribution of J_M, as shown by the following example.

Example:

Consider a scalar system with two regimes

regime 1: $\dot{x} = -x + u$

regime 2: $\dot{x} = u$

and a transition rate matrix $\Pi = \begin{bmatrix} 0 & 0 \\ 0.1 & -0.1 \end{bmatrix}$. Starting from $r_{t_0} = 2$ the regime can only jump to the absorbing regime r = 1 at some exponentially distributed time. The control gains are chosen arbitrarily as

$$\gamma_1 = -2 - \sqrt{7} \qquad \gamma_2 = -1$$

leading to closed-loop dynamics

$$\tilde{a}_1 = -3 - \sqrt{7} \ \ (a_1 = -1) \qquad \tilde{a}_2 = -1 \ \ (a_2 = 0)$$

The first bound in (4.3.1) is evaluated from the value of the suboptimal cost ($x_0 = 1$, $r_{t_0} =$ 2), that is here equal to 2.2/2.1 (from the solution of the cost Liapunov equations):

$$\mathcal{P}\{J_M \geq \lambda\} \leq \frac{2.2}{2.1\lambda}\ .$$

To use Chebyschev's inequality, moments of v are computed as

$$\|v\|_1 = \frac{0.1}{2.05} \left| \frac{2.2}{2.1} - 2 \right| \approx 0.0465$$

and

$$\|v\|_2^2 = \frac{0.1}{4.1} \left| \frac{2.2}{2.1} - 2 \right|^2 \approx 0.0221$$

so that

$$\mathcal{P}\{J_M \geq \lambda\} \leq \frac{0.0221}{(\lambda - 1.118)^2} .$$

In order to evaluate the conservativeness of the bounds the realized cost distribution is also evaluated, as explained in §4.4 below, leading to the exact expression

$$\mathcal{P}\{J_M \geq \lambda\} = 1 - (\lambda - 1)^{0.05}$$

on the limited support set $\lambda \in [1.\ 2.]$. The two bounds are plotted with the exact quantity in figure 4.1.

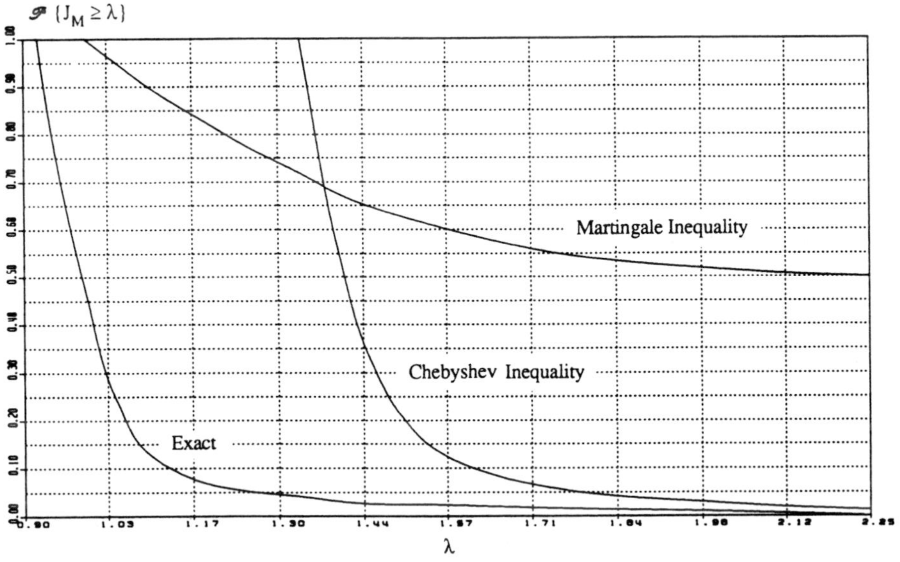

Figure 4.1 - Distribution bounds for the maximal instantaneous cost.

Considering the repeated bounding process leading to (4.3.1), the approximation of the true $\mathcal{P}\{J_M \geq \lambda\}$ is, not surprisingly, rather rough. It does, however, give a preliminary assessment of deviations from the performance predicted by the mean cost at $t = t_0$.

4.4 The Distribution of the Realized Cost

Whereas the variance or bounds on the maximum cost carry partial, though valuable, information on the actual performance, the distribution of the realized cost would provide a complete and accurate characterization. From the development in §4.2 and §4.3 it can be guessed that the computation of this distribution is a very arduous task for general hybrid models and it will not be presented here.

Attention is now given to a rather specialized class of jump systems where the computations become tractable, with the purpose of attaining some insight into the general case.

Let us consider a hybrid system with only two regimes, $r_t = 2$ the transient normal regime with average life-time $1/\pi$ and $r_t = 1$ the trapping failed mode (this is in fact a subcase of the class of models with a directed chain).

The realized cost under $u_t = \Gamma_i x_t$ when $r_t = i$ is given by

$$J_{t_f} = x_0' P(t_0) x_0 \tag{4.4.1}$$

where P is the solution of a stochastic Liapunov equation

$$-\dot{P} = A(r_t)'P + PA(r_t) + Q_1 + \Gamma(r_t)'Q_2\,\Gamma(r_t)$$

$$P(t_f) = 0 \tag{4.4.2}$$

Note that this is the Liapunov counterpart of the stochastic Riccati equation that arose in the maximal principle derivation of the JLQ regulator in chapter 3. Note also that the matrix P in (4.4.1) is related to the P_is, i = 1, N, in chapter 2 by $P_i = E\{P\phi_{ti}\}$. Using the notation $\underline{P}$ for the vector stacking the rows of $\underline{P}$, (4.4.2) can be written

$$\dot{\underline{P}} = F\underline{P} + G$$

$$\underline{P}(t_f) = 0$$

for the regime-dependent matrices F and G

$$F = -(A + B\Gamma) \otimes I_n - I_n \otimes (A + B\Gamma)$$

$$G = \underline{Q_1 + \Gamma' Q_2 \Gamma}$$

The distribution of the realized cost would be known through (4.4.1) if it were possible to compute the distribution of $\underline{P}(t_0)$.

This problem has a simple solution in the present case. Assuming that the distribution of $\underline{P}$ admits a smooth density p(ρ, t, i) ($\mathcal{P}\{\underline{P}(t) \in [\rho, \rho + d\rho] \mid r_t = i\} = p(\rho, t, i)$), it has been shown (Morrison, 1972, Sworder and Choi, 1976a) that p satisfies the following PDE

$$0 = (\text{trace } F_i)\, p(\rho, t, i) + p_\rho(F_i\rho + G_i) + p_t + \sum_{j=1}^{N} \pi_{ij} p\, (\rho, t, j)$$

with $[F_i, G_i] = [F(r_t), G(r_t)]$ when $r_t = i$, and boundary conditions $p(\rho, t_f, i) = \prod_{k=1}^{n^2} \delta(\rho_j)$, 1 to N. The effect of the symmetry of P is that the density does not live on $\mathbf{R}^{n^2}$ but rather on a submanifold of local dimension at most n(n + 1)/2. This unfortunate fact seriously complicates the solution of the density equation and (Sworder and Choi, 1976b) shows the difficulties which arise in the general case. To avoid these difficulties, we shall restrict ourselves to scalar systems (n = 1) where the degeneracy problem disappears. It is also assumed that t_f is large enough for the time dependence to be dropped ($p_t = 0$). When $i_0 = 1$, the trapping regime, the stochastic nature of the problem is degenerate and, from (4.4.2),

$$\mathcal{P}\{\underline{P}(t_0) = -F_1^{-1}G_1\} = 1$$

For $i_0 = 2$, the interesting situation, the characteristic curve C^1 generated by $\dot{\eta} = F_2\eta + G_2$ is introduced. Along C^1, p satisfies

$$\frac{dp}{d\tau} = -(\pi_{22} + F_2)p - \pi_{21}p(\rho, 0, C^1)$$

which is integrated as

$$p(\rho, 2) = \frac{\pi_{21}}{|F_2\underline{P}_1 + G_2|}\left(\frac{F_2\rho + G_2}{F_2\underline{P}_1 + G_2}\right)^{-\frac{\pi_{22}+F_2}{F_2}} \qquad (4.4.3)$$

where $\underline{P}_1 = -F_1^{-1}G_1$.

Equation (4.4.3) gives the explicit density but it remains quite intricate and to display more clearly the role of the various elements, it is transformed back into the usual system parameters. Replacing upper-case symbols by lower-case ones for the present scalar situation (e.g. $\dot{x}_t = a_1x_t + b_1x_t$ when $r_t = 1$) and introducing $\tilde{a}_i = a_i + b_i\gamma_i$, $i = 1, 2$, (4.4.3) can be written

$$p(\rho, 2) = \frac{\pi}{|2\tilde{a}_2(1+\gamma_1^2)/2\tilde{a}_1 - 1 - \gamma_2^2|} \times \left[\frac{-2\tilde{a}_1\rho - 1 - \gamma_2^2}{2\tilde{a}_2(1+\gamma_1^2)/2\tilde{a}_1 - 1 - \gamma_1^2}\right]^{-\frac{2\tilde{a}_2+\pi}{2\tilde{a}_2}} \qquad (4.4.4)$$

where all the cost weighting coefficients are taken equal to 1 and $\pi_{22} = -\pi = -\pi_{21}$ is used. This is a distribution with a bounded support set

$$\left[\mathrm{Min}\left\{\frac{-(1+\gamma_1^2)}{2\tilde{a}_1}, \frac{-(1+\gamma_2^2)}{2\tilde{a}_2}\right\},\ \mathrm{Max}\left\{\frac{-(1+\gamma_1^2)}{2\tilde{a}_1}, \frac{-(1+\gamma_2^2)}{2\tilde{a}_2}\right\}\right] \qquad (4.4.5)$$

The expression obtained in (4.4.4) is admittedly of very limited scope: it gives the distribution of the realized cost only for a scalar system with two regimes, one of which is absorbing. However, its merit is that it is amenable to computer simulation and display. The idea is thus to write a program for the evaluation of (4.4.4) with an easy modification of the model parameters (a_i, b_i, γ_i, $i = 1, 2$), to plot the corresponding curves and to note the qualitative features for different choices of the control gains. Despite the restrictions for performing these computer experiments, the insight they provide opens the way for the next section where new control laws, taking into account the distribution of the realized cost for a large class of hybrid systems, are proposed.

Experiment 1:

The purpose of this experiment is to understand how feedback influences the distribution of the open-loop performance. For $a_1 = -1$, $a_2 = -2$, $b_1 = b_2 = 1$ and $\pi = 1$, theorem 3.1 in chapter 3 gives the gains of the optimal JLQ regulator, called here the Optimal Switching State Feedback (OSSF) as

$$\gamma_1^* = 1 - \sqrt{2}$$

$$\gamma_2^* = \frac{1}{2}\,(5 - \sqrt{25 + 4\sqrt{2}}\,)$$

We recall that throughout these experiments the cost weighting coefficients are kept equal to 1. Entering this data into the program computing (4.4.4) and $\gamma_1 = \gamma_2 = 0$ for the open-loop (OL), figure 4.2 is obtained.

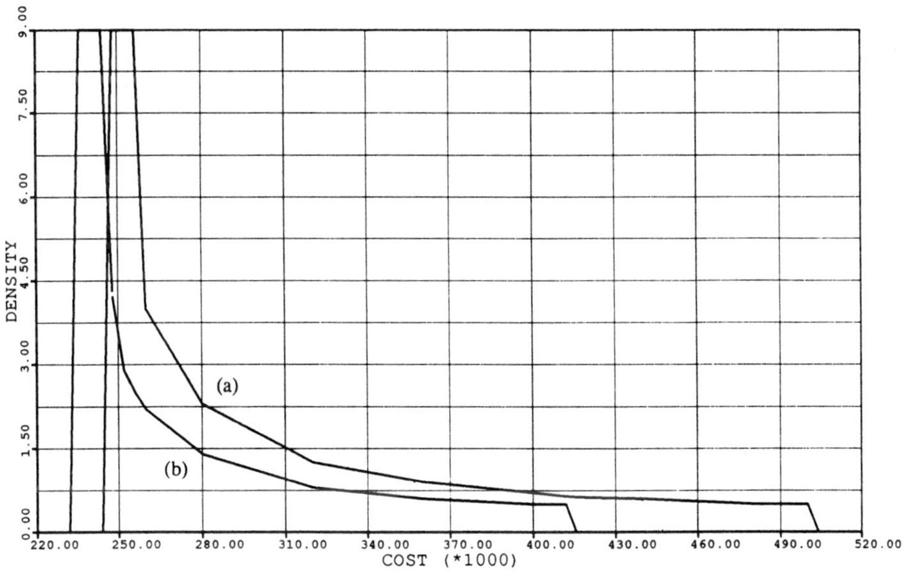

Figure 4.2 - Cost distribution for the Open-Loop system (a) and the system under Optimal Switching State Feedback (b).

As the JLQ regulator optimizes the mean cost, it is natural to observe that the average of the (b) density is shifted to the left of that for the open-loop performance. We observe that the support set is also reduced by feedback.

Experiment 2:

We now wish to observe how a decrease in the mean is translated on the distribution of the realized cost. To do this the OSSF is considered as in experiment 1 and is compared to non-switching feedback laws, that is, with the notations used here $u_t = \gamma x_t, r_t = 1, 2$. For $a_1 = -1, a_2 = 0, b_1 = b_2 = 1$ and $\pi = 1$, the OSSF gains are

$$\gamma_1^* = 1 - \sqrt{2}$$

$$\gamma_2^* = \frac{1}{2}(1 - \sqrt{1 + 4\sqrt{2}})$$

The Optimal Non-Switching State Feedback (ONSSF) is found from theorem 3.6 in chapter 3

$$\gamma^* = -0.722 \quad (\text{for } i_0 = 2)$$

and an Arbitrary Non-Switching State Feedback (ANSSF) is also defined by the gain $\gamma = 1$. Figure 4.3 plots the three corresponding densities.

The first observation concerns the ranking of the mean values, $E_a < E_b < E_c$. This is expected since the optimal solutions seek precisely to minimize the mean and the OSSF solution has the advantage of using more information . But more interesting is the area where the (c) curve is down to zero while (a) still grows, the support of (a) being [0.414, 1.028] and that of (c) [0.5, 1]. This means that while the probability of experiencing a cost greater than 1. under the arbitrary, "non-optimized", control is zero, the mean-optimal OSSF control leads to a cost greater than 1. with a strictly positive probability given by the area of the (a) curve between 1. and 1.028. In other words, there exist sample paths along which the performance under the mean-optimal solution is worse than the worst performance under an arbitrary non-switching solution. This corroborates the example in §4.2 where it was seen that minimization of the mean can result in a significant increase in the variance. This remark is also valid for the class of non-switching solutions where we see that the ONSSF minimizing the mean in fact results in a worse performance than the ANSSF with positive probability.

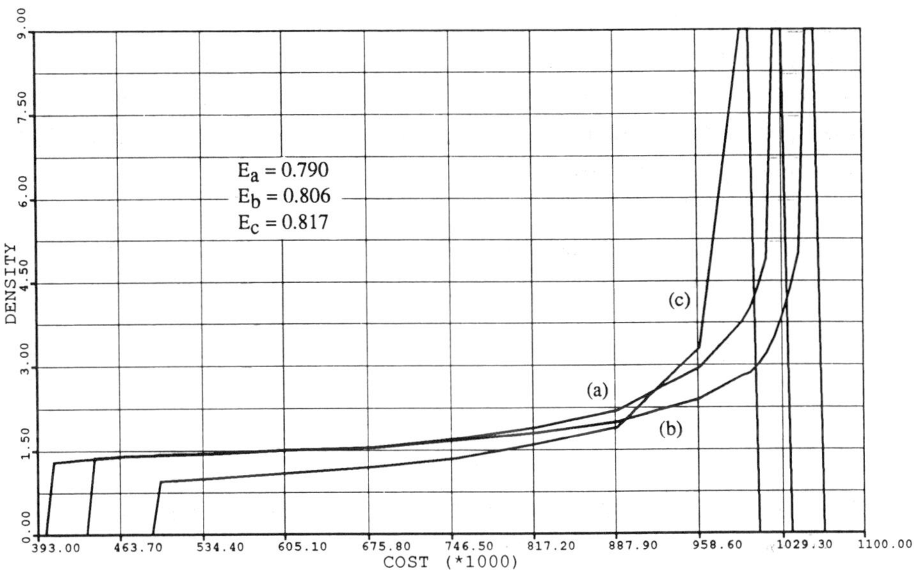

Figure 4.3 - Comparison of the distributions under the Optimal Switching State Feedback (a), the Optimal Non-Switching State Feedback (b) and the Arbitrary Non-Switching State Feedback (c).

As the worst possible performance (that is, the upper bound of the realized cost distribution) is often an important evaluation element, this experiment confirms that the mean cost is not an all-encompassing design criterion and indeed, looking at the densities of figure 4.2, we may even wonder if the (mean-)optimal design really provides the "best" control laws.

This points out the need for a design approach which reflects the intrinsic performance variability introduced by stochasticity.

<u>Experiment 3</u> :

The motivation for this experiment arises from the expression (4.4.5) of the density support set. Obviously if the gains γ_1 and γ_2 are chosen such that

$$\frac{1+\gamma_1^2}{2\tilde{a}_1} = \frac{1+\gamma_2^2}{2\tilde{a}_2}$$

the support set reduces to a single point. The distribution is then degenerated and on any sample path the same performance is obtained. However, the above relation expresses one of the gains as a function of the other and there is in fact a family of solutions with a single point support. Within this class we still have the freedom to look for an optimal solution, that is a single point solution with minimal cost.

For the same parameter values as experiment 2, the "optimal" single point support solution is

$$\gamma_1^{**} = -1 - \sqrt{2}$$

$$\gamma_2^{**} = -1$$

where a double star is used to distinguish from the minimal mean solution. Another single point support set solution is obtained with the gains

$$\gamma_1 = -1$$
$$\gamma_2 = -0.5$$

These two degenerate distributions are plotted on figure 4.4 together with the minimal mean (OSSF) solution.

As with experiment 2, the expected ranking of the means is observed: the OSSF solution, when the average cost is optimized without constraints, produces the smallest mean, in particular $E_a < E_b$ because for (b) optimization is performed under the constraint of a single point support set. However, the degradation of mean performance between (a) and (b) is quite moderate ($\upsilon = E_b - E_a / E_a = 25\%$) and distributions like (b) or even (c) are very attractive from the point of view of performance variability: under (b) or (c) the system behavior, as observed through the cost function, becomes deterministic, meaning that the same performance is obtained along any realization of the jump variable. The realized cost variance is then zero, which seems particularly attractive for the quality control application described in §4.1.

To summarize the outcome of the above experiments, two ideas should be stressed:

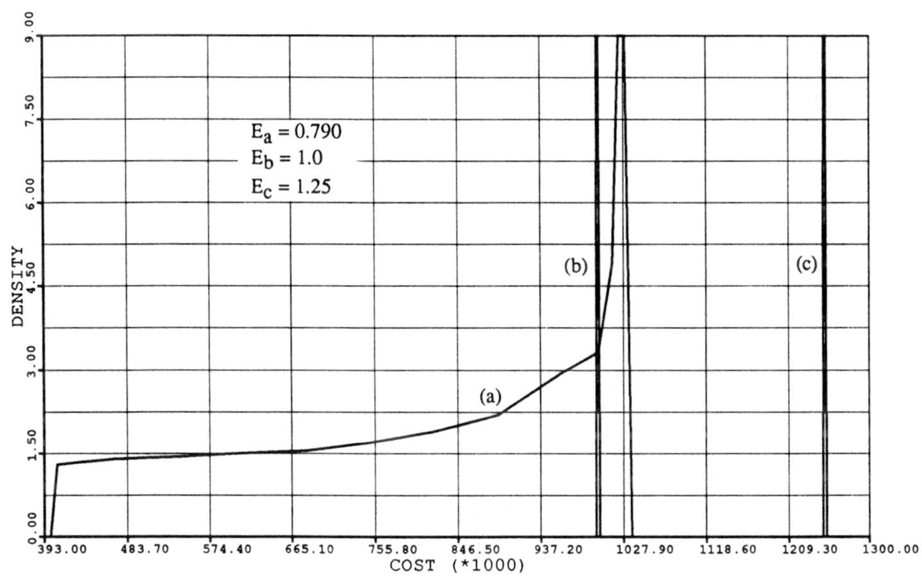

Figure 4.4 - Comparison of the distribution under the minimal mean solution (a) and single point support set optimal (b) and arbitrary (c) solutions.

- Exclusive concern with the mean cost produces regulators that do not shape the realized cost distribution well. Looking a posteriori at measures like the worst possible performance it can be argued that "the optimal solution is not the best one" from the point of view of robustness (experiment 2).
- A new class of solutions was brought to light in experiment 3: the gains can be chosen so that the realized cost distribution becomes concentrated at one point with minimal value. This solution exhibits a remarkable robustness since the realized cost is then a deterministic variable.

Experimenting on a simple specialized class of jump linear systems has thus revealed much of the structure of the realized cost random variable. The most intriguing fact is the expression of the support set bounds and the fact that control design can be aimed at influencing these bounds, collapsing the support set to a single point or minimizing the upper bound.

Building on these observations, we now wish to understand how to generalize the above findings to a more general (non scalar, several regimes) hybrid model. By interpreting stochastic control as a game against Nature, the answer is found in §4.5.

4.5 Equalizing and Minimax Solutions

The outcome of a poker game depends not only on the players' decisions but also on the way cards are dealt in the first place. This chance move can be attributed to Nature acting as an additional player, and the purpose of statistical games theory is then to understand and define rational ways in which a human player, called the statistician, should behave when faced with the randomness of Nature's behavior. As stated, the scope of this theory is rather wide and covers the numerous situations where decisions have to be taken in the presence of uncertainty. Early contributions can be traced back to E. Borel, but it is Wald, exploiting the connections with Von Neumann's emerging theory of games, who really laid the foundations of the modern decision theoretic viewpoint with contributions such as his celebrated sequential probability ratio test.

Replacing the "statistician" by the more engineering-like "designer", it is clear that the formulation of statistical games theory encompasses many aspects of stochastic control. The designer is asked to propose a controller to be implemented on a plant whose behavior is in turn influenced by random perturbation signals: while the designer chooses the control law to shape the realized cost distribution, it is Nature which ultimately decides the sample path performance by picking up ("drawing") a realization of the underlying random processes.

Two key concepts immediately relevant to the design of controllers for stochastic systems are the concepts of equalizing and minimax solutions.

The minimax idea is to look for the decision that minimizes the maximal value of the cost that Nature may choose in the most adverse situation. In terms of the densities discussed in §4.4, this solution minimizes the upper-bound of the support set. The attractiveness of the minimax strategies is that they guarantee an upper-bound on the cost (or a lower-bound on the performance) without regard to the actual, sample path, evolution of the state dynamics as decided by Nature.

The equalizing concept corresponds to the class of solutions leading to degenerate distributions with single point support which appeared in example 2 of the previous section. As the name and figure 4.4 suggest, an equalizer makes the performance independent of the choice of Nature: on any realization the same performance is obtained and the cost thus becomes a deterministic variable.

Motivated by the examples in §4.4 for $n = 1$, it seems that the worst and the best choice of Nature for the JLQ should correspond to constant, no jump, regime realizations.

To prove this fact, a JL system with two regimes $r = 1$ trapping and $r = 2$ transient, is considered first. A related attempt, using a different technique, was reported in (Sworder, 1977).

From (4.4.4) the value of the realized cost is

$$J = \underline{P}(t_0)'X_0 \tag{4.5.1}$$

where X_0 is the n^2 column vector stacking the rows of the nxn matrix x_0x_0'. We recall that $\underline{P}$ satisfies

$$\dot{\underline{P}} = F(r_t)\,\underline{P} + G(r_t)\,; \qquad \underline{P}(t_f) = 0 \tag{4.5.2}$$

for some random F and G which were defined between (4.4.2) and (4.4.3). When $N = 2$, only one jump is possible, a jump at, say, $t = t_{21}$, from $r = 2$ to $r = 1$. The choice of Nature in this special case is thus restricted to either playing regime 1 or regime 2 throughout the game ($r_t = 1$ or 2 from t_0 to t_f) or forcing the regime to be 2 from t_0 to t_{21} and 1 after that. The first term of the alternative corresponds to the property observed in §4.4, and our objective is therefore to show that the bounds of the distribution support are attained for these constant choices of Nature.

To do this it is sufficient to prove that J, considered as a function of t_{21}, has no extremum on $]t_0, t_f[$. We integrate (4.5.2) from t_f to t_{21} to obtain

$$\underline{P}(t_{21}) = F_1^{-1}(e^{F_1(t_{21}-t_f)} - I)\,G_1 \tag{4.5.3}$$

and then from t_{21} to t_0

$$\underline{P}(t_0) = \underline{P}(t_{21}) + F_2^{-1}(e^{F_2(t_0-t_{21})} - I)\,(G_2 + F_2\underline{P}(t_{21})) \tag{4.5.4}$$

where F_1 and F_2 are assumed to be regular.

We now consider the asymptotic case ($t_f \to \infty$) and, recalling that F_1 and F_2 have all their eigenvalues unstable, (4.5.1), (4.5.3) and (4.5.4) yield

$$J = [F_2^{-1}e^{-F_2t_{21}}(G_2 - F_2F_1^{-1}G_1) - F_2^{-1}G_2]'\,X_0$$

where t_0 has been temporarily set to zero to simplify notation. Upon differentiating,

$$\partial J/\partial t_{21} = [F_1^{-1}G_1 - F_2^{-1}G_2]'\,F_2'e^{-F_2't_{21}}\,X_0$$

Observe that for $t_{21} = \infty$ (resp. $t_{21} = 0$), $J = -\,[F_2^{-1}G_2]'\,X_0$ (resp. $J = -\,[F_1^{-1}G_1]'\,X_0$) which is the LQ cost associated with the operation under regime 2 (resp. regime 1) from $t = 0$ to $t = \infty$. These costs give an ordering between regime 1 and regime 2 that is

$$-\,[F_2^{-1}G_2]'\,X_0 \le \text{ or } \ge -\,[F_1^{-1}G_1]'\,X_0 \tag{4.5.5}$$

This ordering depends in general on the initial condition, but attention is now restricted to the class of systems for which one of the inequalities holds for any X_0

$$\begin{array}{ll} -\,[F_2^{-1}G_2]'\,X_0 \le -\,[F_1^{-1}G_1]'\,X_0 & \\ \text{or} & \forall\, X_0 \\ -\,[F_2^{-1}G_2]'\,X_0 \ge -\,[F_1^{-1}G_1]'\,X_0 & \end{array} \tag{4.5.6}$$

Though restrictive, this assumption is quite reasonable: thinking, for example, of the fault-tolerant control application and associating regime 2 with the nominal regime and regime 1 with degraded operation, (4.5.6) means that regime 2 is superior to regime 1 (the first inequality in (4.5.5)) uniformly in X_0. In other words, the system always does better in the nominal regime than in a failed regime. This condition is satisfied for regime-independent weights in (4.1.1), when failures uniformly move the eigenvalues of $A(r_t)$ to the right with $B(r_t)$ constant or reduce the control influence matrix, that is, $B(r_t)B(r_t)'$, with $A(r_t)$ constant.

Within this class of systems, consider the case

$$- [F_2^{-1}G_2]' X_0 \leq - [F_1^{-1}G_1]' X_0 \qquad (4.5.7)$$

We have

$$\partial J/\partial t_{21} = [F_1^{-1}G_1 - F_2^{-1}G_2]' \bar{X}_0 \qquad (4.5.8)$$

for $\bar{X}_0 = F_2'e^{-F_2't_{21}} X_0$. As (4.5.7) holds for any X_0, especially $\bar{X}_0$, we obtain

$$\partial J/\partial t_{21} \leq 0 \qquad \forall\ t_{21} \in [0, \infty[$$

From this we conclude that J decreases monotonically from $- [F_1^{-1}G_1]' X_0$ for $t_{21} = 0$ to $- [F_2^{-1}G_2]' X_0$ for $t_{21} = \infty$. Hence the extrema of the cost with respect to the choice of Nature (t_{21}) occur at $t_{21} = 0$ (J maximal) and $t_{21} = \infty$ (J minimal); that is, on constant regime trajectories $r_t = 2$ and $r_t = 1$ over $[0, \infty[$. The same reasoning gives a monotonically increasing cost when the inequality is reversed in (4.5.7).

In the general case N > 2, the counterpart of (4.5.7) is

$\exists\ i_w$ and $i_b \in$ **S** such that

$$\left.\begin{array}{l} - [F_{i_w}^{-1}G_{i_w}]' X_0 \geq - [F_i^{-1}G_i]' X_0 \\ - [F_{i_b}^{-1}G_{i_b}]' X_0 \leq - [F_i^{-1}G_i]' X_0 \end{array}\right. \qquad \forall\ i \in \mathbf{S},\ \forall\ X_0 \qquad (4.5.9)$$

which states that there exists a regime i_w (resp. i_b) giving the worst (resp. best) LQ performance uniformly in X_0. The following theorem is then stated.

<u>Theorem 4.3</u>

For the JLQ problem with condition (4.5.9) satisfied, a directed chain for the regime, and under a linear switching state feedback, the worst (resp. best) choice of Nature is a constant realization of the regime process with $r_t = i_w$ (resp. $r_t = i_b$) from t_0 to $t_f \to \infty$, where i_w (resp. i_b) is the regime with the largest (resp. smallest) LQ cost.

The LQ costs are $x_0'P_i^1x_0$ (= $-[F_i^{-1}G_i]'X_0$), i = 1, N with matrice P_i^1 as given by the N Liapunov equations

$$0 = \tilde{A}_i' P_i^1 + P_i^1\tilde{A}_i + Q_i + \Gamma_i'R_i\Gamma_i \qquad i = 1 \text{ to } N \tag{4.5.10}$$

with $\tilde{A}_i = A_i + B_i\Gamma_i$.

Proof:

The idea is again to show that the cost is a monotone function of the jump instants when (4.5.9) is satisfied. The computations for N = 3 illustrate the general mechanism of the proof while limiting complexity in the notation, and thus only this case is given below.

For N = 3, Nature's choices are the instants t_{32} and t_{21} when r_t jumps from 3 to 2 (t_{32}) and from 2 to 1 (t_{21}). From (4.5.2), and recalling that $t_f = \infty$, we obtain

$$\underline{P}(t_{21}) = -F_1^{-1}G_1$$

$$\underline{P}(t_{32}) = F_2^{-1}e^{F_2(t_{32}-t_{21})}(G_2 - F_2F_1^{-1}G_1) - F_2^{-1}G_2$$

$$\underline{P}(t_0) = F_3^{-1}e^{F_3(t_0-t_{32})}(G_3 - F_3\underline{P}(t_{32})) - F_3^{-1}G_3$$

with J given by (4.5.1). It can be observed that

$$\begin{aligned}
&\text{for } t_{21} = t_{32} = \infty && J = -[F_3^{-1}G_3]'X_0 \\
&\text{for } t_{21} = \infty \text{ and } t_{32} = 0 && J = -[F_2^{-1}G_2]'X_0 \\
&\text{for } t_{21} = t_{32} = 0 && J = -[F_1^{-1}G_1]'X_0
\end{aligned} \tag{4.5.11}$$

That is, the LQ costs associated with constant regime trajectories, respectively $r_t = 3, 2$ and 1 over $[t_0, \infty[$. Condition (4.5.9) then states that the ordering of the three costs in (4.5.10) is independent of X_0, say

$$-[F_3^{-1}G_3]'X_0 \leq -[F_2^{-1}G_2]'X_0 \leq -[F_1^{-1}G_1]'X_0.$$

Using steps like (4.5.8), we then see that

$$\partial J/\partial t_{21} \le 0 \text{ and } \partial J/\partial t_{32} \le 0 \qquad \text{for } t_0 \le t_{32} \le t_{21} \le \infty$$

so that J decreases monotonically from - $[F_1^{-1}G_1]' X_0$ for $t_{21} = t_{32} = 0$ to - $[F_3^{-1}G_3]' X_0$ as t_{21} and t_{32} are increased. □

Using the best and worst regimes of the above theorem, we shall now discuss equalizing and minimax solutions.

Equalizing solutions

The above theorem reveals precisely the way the bounds of the cost distribution support are generated. It is then easy to see how the "single point support" solutions from example 2 in §4.4, now called equalizing from game theory, can be generalized: if the feedback gains Γ_i, i = 1 to N, in (4.5.10) are chosen such that the LQ costs $x_0'P_i^1x_0$, i = 1 to N, are all equal to J_{eq}, the distribution degenerates to $\mathscr{P}\{J = J_{eq}\}$ = 1.

Under control, the mean value of the realized cost is given by

$$J = x_0'P_{i_0}^0x_0 \qquad \text{when } r_{t_0} = i_0$$

where the P_i^0s, I = 1 to N, are solutions of a set of coupled Liapunov equations

$$0 = \tilde{A}_i' P_i^0 + P_i^0\tilde{A}_i + \Gamma_i'R_i\Gamma_i + Q_i + \sum_{j=1}^{N} \pi_{ij} P_j^0 \qquad i = 1 \text{ to } N \qquad (4.5.12)$$

It was proved in (Sworder, 1977) that for a JL system with lower triangular Π the variance of the cost under the mean optimal policy is increased every time a jump occurs. Note that the above proof and (Sworder, 1977) concur in that from Sworder trajectories where the mode does not jump contribute nothing to the variance.

From example 2 in §4.4 it is clear that there are in general many equalizing solutions. With the notations of this example it is seen that any γ_2 solution of

$$r_2\gamma_2^2 - b_2(q_1 + \gamma_1^2r_1)/(a_1 + b_1\gamma_1)\gamma_2 + q_2 - a_2(q_1 + \gamma_1^2r_1)/(a_1 + b_1\gamma_1) = 0$$

(with $\tilde{a}_1$ and $\tilde{a}_2$ stable and γ_2 fixed) provides an equalizer.

A legitimate desire is then to use the remaining degrees of freedom to achieve other objectives besides equalization. The cost function reflects the regulation performance that the designer wants to achieve and it is thus reasonable to look for an optimal equalizing solution that would minimize this cost.

As the bounds of the distribution support correspond to the LQ costs, the associated optimal LQ regulators are introduced

$$u_t = - Q_{2i}^{-1} B_i' \Lambda_i^1 x_t \qquad \text{when } r_t = 1$$

where the Λ_i^1 matrices, i = 1 to N, are the solutions of N decoupled Riccati equations

$$0 = A_i' \Lambda_i^1 + \Lambda_i^1 A_i - \Lambda_i^1 B_i Q_{2i}^{-1} B_i' \Lambda_i^1 + Q_{1i} \qquad i = 1 \text{ to } N$$

The value $x_0'\Lambda_i^1 x_0$ is then the smallest possible value of the ith LQ cost. Computing the Λ_i^1, i = 1 to N, therefore provides the desired ordering of the regimes for a given x_0, with i_w giving the largest value and i_b the smallest value of $x_0'\Lambda_i^1 x_0$.

The minimal value of the equalized cost J_{eq} is then obviously

$$J_{eq}^{**} = x_0'\Lambda_{i_w}^1 x_0$$

since any lower value could not be attained by the i_wth regime. For the N - 1 other regimes, equalizing means increasing their LQ costs by moving the gains away from their optimal value. The LQ cost $x_0'P_i^1 x_0$ being an unbounded continuous function of the gain Γ_i, there exists a gain such that $x_0'P_i^1 x_0 = J_{eq}^{**}$ ($\geq x_0'\Lambda_i^1 x_0$) and a gradient algorithm converging to such a gain is proposed below.

This discussion has proved the following corollary.

Corollary 1

When considering the JLQ problem, the minimal cost for an equalizing solution under condition (4.5.9) is given by the worst optimal LQ cost

$$J_{eq}^{**} = x_0'\Lambda_{i_w}^1 x_0 \geq x_0'\Lambda_i^1 x_0 , \qquad i = 1 \text{ to } N, i \neq i_w$$

It can be observed that for the $i \neq i_w$ regimes the optimal equalizer implies a degradation of their individual performance. This inconvenience has to be balanced against the decoupling of the J_{eq}^{**} performance with the choice of Nature and, if the cost increase remains moderate (as in example 2 of §4.4 above), the designer will often select the optimal equalizing strategy.

Optimal equalizing gains Γ_i^{**}, i = 1 to N, corresponding to J_{eq}^{**}, are next obtained. One of these is already known, namely the i_wth one, equal to the optimal LQ gain. For the others there are in fact N - 1 separate identical problems and the indices are temporarily dropped, A, B, Γ^{**},..., designating one of the A_i, B_i, Γ_i^{**}, ..., i = 1 to N, $i \neq i_w$. To ease the presentation of the algorithm, x_0 is assumed to be a centered random variable with $E\{x_0x_0'\} = I_n$.

At the initial step, find a stabilizing gain $\Gamma^{(0)}$ close to the optimal LQ gain, for example

$$\Gamma^{(0)} = - (1 + \varepsilon) Q_2^{-1}B' \Lambda^1$$

or

$$\Gamma^{(0)} = - (1 - \varepsilon) Q_2^{-1} B' \Lambda^1$$

with $\varepsilon \sim 0.1$.

The gradient of the LQ cost (= trace P^1) is

$$\partial J/\partial\Gamma = Q_2\Gamma + B'P^1$$

and for $\Gamma = \Gamma^{(c)}$, $P^1 = P^{1(c)}$, the solution of (4.5.10), there exists a step size $\alpha^{(c)} \in \mathbf{R}_+^*$ such that the ascent

$$\Gamma^{(c+1)} = \Gamma^{(c)} + \alpha^{(c)} (\partial J/\partial\Gamma)^{(c)}$$

preserves stability ($A + B\Gamma^{(c+1)}$ stable) and increases the cost

$$\text{trace } P^{1(c+1)} \geq \text{trace } P^{1(c)}.$$

A c_0 therefore exists such that

$$\text{trace } P^{1(c_0+1)} \geq \text{trace } \Lambda^1_{i_w} \geq \text{trace } P^{1(c_0)}$$

(if a local extremum stops the algorithm before a c_0 is encountered, some reversed steps of the gradient are performed).

Then restart the gradient ascent at $\Gamma = \Gamma^{(c_0)}$ with a reduced step size (for example $\alpha^{(c_0+1)} = 0.2\ \alpha^{(c_0)}$. Iterating the above steps will give

$$\lim_{c \to \infty} \Gamma^{(c)} = \Gamma^{**}$$

such that trace $P^1(\Gamma^{**})$ comes arbitrarily close to trace $\Lambda^1_{i_w}$.

Remark:

The initial gradient near the global minimum will not be computed with great accuracy but this is not serious since all that is desired is to increase the cost while preserving stability.

There is of course no claim that there exists a unique set of gains Γ^{**}_i, i = 1 to N. The gain in the worst case Γ_{i_w} being fixed, it is intuitive that the other LQ costs can be increased by either "increasing" or "decreasing" the matrix gains $\Gamma_{i \neq i_w}$: J is then increased through either the control penalization (larger gains) or the state penalization (smaller gains). The above algorithm can be oriented towards one or the other possibility at the initialization step by choosing the $(1 + \varepsilon)$ or $(1 - \varepsilon)$ multiplier. This choice is left to the designer, who will base it on his knowledge of the system peculiarities, as discussed by (Sworder, 1977). In the case of economic planning a small gain solution would certainly be preferred to take into account the political constraints that surround any policy adjustment.

Minimax solutions

From theorem 4.3 it appears that the upper bound of the distribution support is associated to the LQ cost for the worst case regime i_w. The minimal value of this upper bound is therefore obtained when

$$\Gamma_{i_w} = - Q_{2i}^{-1} B_{i_w}' \Lambda_{i_w}^1 \quad (= \Gamma_{i_w}^{**}) \tag{4.5.13}$$

with the notations of (4.5.9) where i_w is the same as in Corollary 1. Note that the gain when $r_t = i_w$ is the same as the optimal equalizing gain. It is again clear that the remaining N - 1 gains are free, provided they do not lead to an LQ cost larger than the i_wth optimal LQ one. These degrees of freedom are exploited next.

As for the optimal equalizer, a logical way to exploit the remaining degrees of freedom of a minimax solution is to try to minimize the average value of the cost function (4.1.2) which still reflects a design objective. Under a set of gains Γ_i, i = 1 to N, $i \neq i_w$ and Γ_{i_w} given by (4.5.13), the average value of the cost is

$$E\{J\} = x_0' P_{i_0}^2 x_0 \qquad \text{when } r(t_0) = i_0$$

with the P_i^2, i = 1 to N, given as

$$0 = \tilde{A}_i' P_i^2 + P_i^2 \tilde{A}_i + \Gamma_i' Q_{2i} \Gamma_i + Q_{1i} + \sum_{j=1}^{N} \pi_{ij} P_j^2 \qquad i = 1 \text{ to } N \tag{4.5.14}$$

Noting that (4.5.14) is just a set of coupled Liapunov equations, the minimization is obtained as usual by setting

$$\Gamma_i = - Q_{2i}^{-1} B_i' P_i^2 \qquad i = 1 \text{ to } N, i \neq i_w$$

in (4.5.14). However, this is now possible only for $i \neq i_w$ since the i_w^{th} gain is fixed by (4.5.13). We thus define the Λ_i^2, i = 1 to N, as

$$0 = A_i' \Lambda_i^2 + \Lambda_i^2 A_i - \Lambda_i^2 B_i Q_{2i}^{-1} B_i' \Lambda_i^2 + Q_{1i} + \sum_{j=1}^{N} \pi_{ij} \Lambda_j^2 \qquad i \neq i_w$$

$$0 = \tilde{A}_{i_w}' \Lambda_{i_w}^2 + \Lambda_{i_w}^2 \tilde{A}_{i_w} + \Lambda_{i_w}^1 B_{i_w} Q_{2i_w}^{-1} B_{i_w}' \Lambda_{i_w}^1 + Q_{1i_w} + \sum_{j=1}^{N} \pi_{i_w j} \Lambda_j^2 \quad i = 1 \text{ to } N \tag{4.5.15}$$

Note that since the P_i^2s define the expected value of the cost and $\Lambda_{i_w}^1$ the upper bound of its distribution support we have

$$x_0'P_i^2\, x_0 \le x_0'\Lambda_{i_w}^1\, x_0 \qquad\qquad i = 1 \text{ to } N$$

and, because the Λ_i^2 result from the minimization of the cost

$$x_0'\Lambda_i^2\, x_0 \le x_0'P_i^2\, x_0 \le x_0'\Lambda_{i_w}^1\, x_0 \qquad i = 1 \text{ to } N$$

The gains of the optimal minimax solutions

$$\Gamma^{*\,*\,*}_{i} = - Q_{2i}^{-1}B_i'\,\Lambda_i^2 \qquad\qquad i = 1 \text{ to } N, i \ne i_w \qquad (4.5.16)$$

with $\Gamma^{*\,*\,*}_{i_w}$ given by (4.5.13) therefore do not change the ordering of the modes, and the worst choice of Nature is still the same i_w.

This discussion has proved the following corollary.

Corollary 2

The gains of the optimal minimax solution for the JLQ problem under condition (4.5.9) are given by (4.5.13), (4.5.15) and (4.5.16). The upper bound of the distribution support (the minimax cost) is

$$x_0'\Lambda_{i_w}^1\, x_0$$

with the notations of (4.5.9) and Corollary 1, and the minimal average value of the cost is

$$x_0'\Lambda_{i_0}^2\, x_0 \qquad\qquad \text{when } r(t_0) = i_0$$

Note that the solutions obtained so far involve design equations of unequal complexity. The JLQ mean optimal regulator requires the solution of N coupled Riccati equations. An equalizer is obtained more easily since only the N decoupled Liapunov equations (4.5.10) are used. However, if an optimal equalizer is desired the complexity is increased with the N decoupled Riccati equations for the Λ_i^1, i = 1 to N, and the algorithm iterations for the L^{**}_i, $i \ne i_w$. Similarly a minimax solution is produced with only N decoupled Riccati equations, but adding the minimal mean requirement leads to the additional coupled (N - 1 Riccati, 1 Liapunov) equations (4.5.15).

This suggests a remark regarding the robustness of the equalizing and minimax solutions with respect to the entries of the transition rate matrix Π. It can be observed that the decoupled set of equations are <u>independent</u> of Π. This provides an exceptional robustness which is interesting in practice when there is often a significant range of uncertainty surrounding the entries of Π. Independence on Π is preserved for the optimal equalizer, but unfortunately optimization of the minimax solution results in design equations (4.5.15) that again depend on Π (just as did the mean optimal JLQ regulator) and thus destroy this property.

Finally, the properties of the optimal minimax solutions compared to the mean optimal JLQ regulator are illustrated by considering again an example similar to those in §4.4.

<u>Example:</u>

With the usual notation, consider

$$a_1 = 0, \quad a_2 = 5, \quad b_1 = 1, \quad b_2 = 1,$$
$$\pi = 10,$$
$$q_{11} = q_{12} = q_{21} = q_{22} = 1.$$

Regime 2, which is open-loop unstable in a deterministic sense ($a_2 > 0$) has a short average lifetime (0.1 time unit) and r_t then jumps in the more favorable trapping mode 1.

The mean optimal solution given in chapter 3 is

$$\gamma_1^* = -1, \qquad \gamma_2^* = -\sqrt{11},$$

and the optimal minimax solution (4.5.16)

$$\gamma_1^{**} = -1 \qquad \gamma_2^{**} = -5 - \sqrt{26}$$

The mean optimal solution does not, in a deterministic sense, stabilize the transient regime ($\tilde{a}_2^* = 5 - \sqrt{11} > 0$) because it has a short average lifetime and therefore contributes little once the expectation of the cost is taken. On the other hand the minimax solution takes into account the worst possible choice of Nature, which here is to play regime 2 throughout the game and therefore takes care to stabilize this regime ($\tilde{a}_2^{**} = -\sqrt{26} < 0$). This of course requires more control expenditure and results in a degraded average performance. If on a given sample path regime 2 persists for a significant length of time, the JLQ regulator, exclusively concerned with the mean performance, produces a diverging state trajectory. In such a situation the minimax regulator appears as the strategy to be chosen.

4.6. A Robust JLQ Regulator

We now turn to a different aspect of robust design and propose a modification of the JLQ regulator that guarantees regime deterministic stability. We drop the directed chain assumption used from §4.2 to §4.5.

Theorems 3.2 and 3.3 in chapter 3 studied stabilization properties of the asymptotic JLQ regulator. Theorem 3.2 gave sufficient conditions for the matrices $A_i + B_i\Gamma_i + \frac{1}{2}\pi_{ii} I_n$, i = 1 to N, to be stable in a deterministic sense if the Γ_is are chosen as the optimal JLQ gains. However, as π_{ii} is negative, this does not imply stability of the closed-loop dynamics $\tilde{A}_i = A_i + B_i\Gamma_i$ but only that $\sigma(\tilde{A}_i) < -\frac{1}{2}\pi_{ii}$. For regimes with short averaged lifetimes ($-1/\pi_{ii}$ small), unstable dynamics ($\sigma(\tilde{A}_i) > 0$) can be obtained as a result of JLQ optimization (see the examples in chapter 3). The mechanism leading to this phenomenon was explained in chapter 2 by the interactions between the discrete and continuous components of the hybrid system and this was captured in the notion of averaged dynamics. Theorem 3.3 is then the most precise result available where necessary and sufficient conditions are given for the asymptotic JLQ regulator to produce stable mean-square dynamics, a stochastic form of stability involving the $\tilde{A}_i$s, i = 1, N coupled through regime jumps.

The disturbing fact is therefore that the basic JLQ regulator fails to stabilize the individual regimes in a deterministic sense ($\sigma(\tilde{A}_i) < 0$, i = 1, N). This failure makes the JLQ regulator not robust with respect to the uncertainty surrounding the realization of the underlying stochastic process along a sample path: typically $|\pi_{ii}|$ large only implies that the *average* lifetime of regime i is short, and when the process runs it may well happen that it stays much longer than $-1/\pi_{ii}$ in that regime. An instability of $\tilde{A}_i$ may then produce catastrophic damage to the system, for example when the state diverges in a transient degraded regime. It should also not be forgotten that the JLQ design is based on linearized state dynamics which are valid only in a limited neighborhood of the equilibrium. A growth of x_t when $\tilde{A}_i$ is unstable may then lead the process out of this region, rendering the entire JLQ solution useless and making recovery difficult.

On the other hand, a robust regulator should stabilize the $\tilde{A}_i$s individually to guard the system against longer than average stays in some regimes along a sample path.

The almost sure stability result of chapter 2 provides an answer to this request: requiring deterministic stabilizability of the $[A_i, B_i]$ pairs, i = 1, N, it shows that choosing the gains Γ_i to have $\sigma(\tilde{A}_i) < 0$, i = 1, N is sufficient to ensure almost sure stability. However, such a design would ignore the couplings between regimes induced by jumps and it could result in a poor performance, as measured by the averaged JLQ cost.

Preserving the cost optimization design and satisfying the robustness requirements of stable regimes is nevertheless possible through an adaptation of the JLQ formulation, as is now shown.

Instead of the cost used in chapter 3, consider the modification

$$J_m = E\{ \int_{t_0}^{t_f} e^{\theta t}(x'Q_1x + u'Q_2u)\, dt \mid x_0, i_0, t_0\} \tag{4.6.1}$$

where θ acts as a discount rate.

The JLQ regulator derivations applied to (4.6.1) lead to the modified set of Riccati equations

$$-\dot{\Lambda}_i = (A_i + \frac{1}{2}\theta\, I_n)'\,\Lambda_i + \Lambda_i\,(A_i + \frac{1}{2}\theta\, I_n) - \Lambda_i B_i Q_{2i}^{-1} B_i' \Lambda_i + Q_{1i} + \sum_{j=1}^{N} \pi_{ij}\Lambda_j \tag{4.6.2}$$

$$\Lambda_i\,(t_f) = 0$$

where it is apparent that a positive θ can be used to force the regulator to counter the influence of $\frac{1}{2}\,\pi_{ii}$ in $\tilde{A}_i + \frac{1}{2}\,\pi_{ii}I_n$. This is shown below.

Theorem 4.4

Suppose that the triplets $[A_i + \frac{1}{2}\theta\, I_n, B_i, Q_{1i}{}^{1/2}]$, i = 1 to N, are stabilizable and detectable in the usual deterministic sense, and that the regime generator is such that

$$\underset{i=1,N}{\text{Max}}\ \underset{\Gamma}{\text{Inf}} \left| \pi_{ii} \int_0^\infty \exp(A_i + B_i\Gamma + \frac{1}{2}(\pi_{ii} + \theta)I_n)'t \exp(A_i + B_i\Gamma + \frac{1}{2}(\pi_{ii} + \theta)I_n)t\, dt \right| < 1$$

The Λ_i, i = 1 to N, solutions of (4.6.2) then converge when $t_0 \to -\infty$ towards the $\bar{\Lambda}_i$, i = 1 to N, unique positive semi-definite solutions of

$$0 = (A_i + \frac{1}{2}\theta\, I_n)'\bar{\Lambda}_i + \bar{\Lambda}_i(A_i + \frac{1}{2}\theta\, I_n) - \bar{\Lambda}_i B_i Q_{2i}^{-1} B_i'\bar{\Lambda}_i + Q_{1i} + \sum_{j=1}^{N} \pi_{ij}\bar{\Lambda}_j \tag{4.6.3}$$

and for the choice $\theta = \underset{i=1,N}{\text{Max}} |\pi_{ii}|$ of the discount rate the corresponding closed-loop system is stable in a deterministic sense ($\sigma(\tilde{A}_i) < 0$, i = 1 to N).

Proof:

It is a repetition of the proof of theorem 3.2 in chapter 3 adapted to the Riccati equations (4.6.2). The stabilization is now obtained for the matrices $\tilde{A}_i + \frac{1}{2}(\theta + \pi_{ii})\, I_n$, i = 1 to N, and the choice $\theta = \underset{i=1,N}{\text{Max}} |\pi_{ii}|$ is made to have $\theta + \pi_{ii} \geq 0$, $\forall$ i = 1 to N, so that the stability of $\tilde{A}_i + \frac{1}{2}(\theta + \pi_{ii})\, I_n$ implies $\sigma(\tilde{A}_i) < 0$, i = 1 to N. □

The modified cost (4.6.1) therefore tackles the lack of robustness of the mean optimal design and the regulator $u_t = -Q_{2i}^{-1}B_i\bar{\Lambda}_i x_t$ when $r_t = i$ for the $\bar{\Lambda}_i$s given by (4.6.3) is referred to as the robust JLQ regulator.

To illustrate this aspect of robustness, consider a scalar model with two regimes with $a_1 = -1$, $a_2 = 1$, $b_1 = b_2 = 1$ and $\pi_{11} = 0$, $\pi_{22} = -2\pi$. With equal weighting coefficients the JLQ regulator is computed and the corresponding closed-loop dynamics in regime 2 are

$$\tilde{a}_2 = \pi - \sqrt{(\pi-1)^2 + 1 + 2\pi(\sqrt{2}-1)}$$

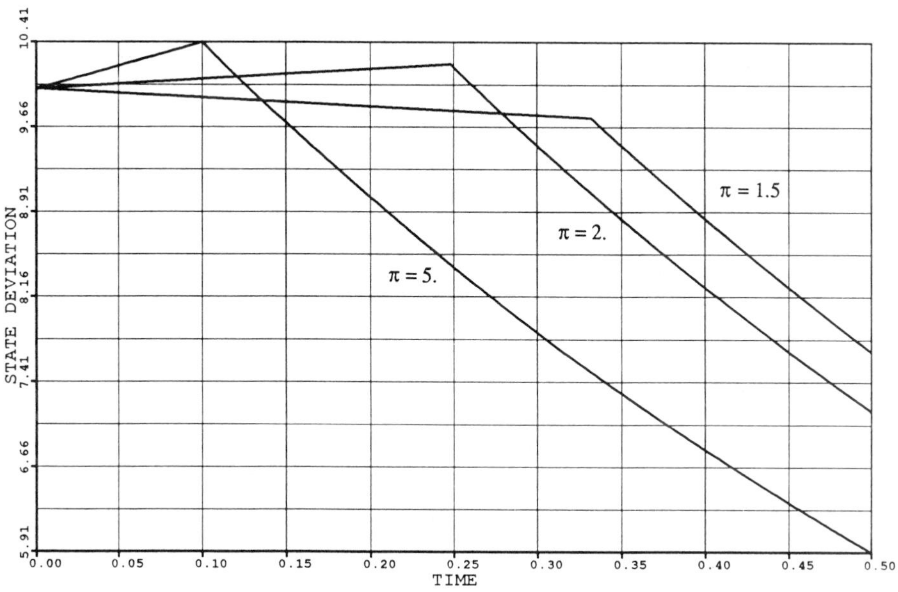

Figure 4.5 - Existence of unstable controlled dynamics.

which is unstable when regime 2 has a short lifetime ($\tilde{a}_2 > 0$ for $\pi \geq 1/(2-\sqrt{2})$). Figure 4.5 plots trajectories with jump at $t=1/2\pi$ for several values of π.

For example, for $\pi = 5$, x_t grows until r_t jumps to 1 and on sample paths where the sojourn time in $r_t = 2$ is significantly longer than 1/10 time units a very large value of x_t will result. Clearly this solution does not meet reasonable robustness requirements for large deviations from average regime realizations.

On the other hand the robust JLQ regulator of theorem 4.4 leads to

$$\tilde{a}_2 = -\sqrt{(\pi-1)^2 + 1 + 2\pi(\sqrt{2}-1)}$$

which is stable for any value of π. Average trajectories under the robust and non-robust regulators are compared in figure 4.6.

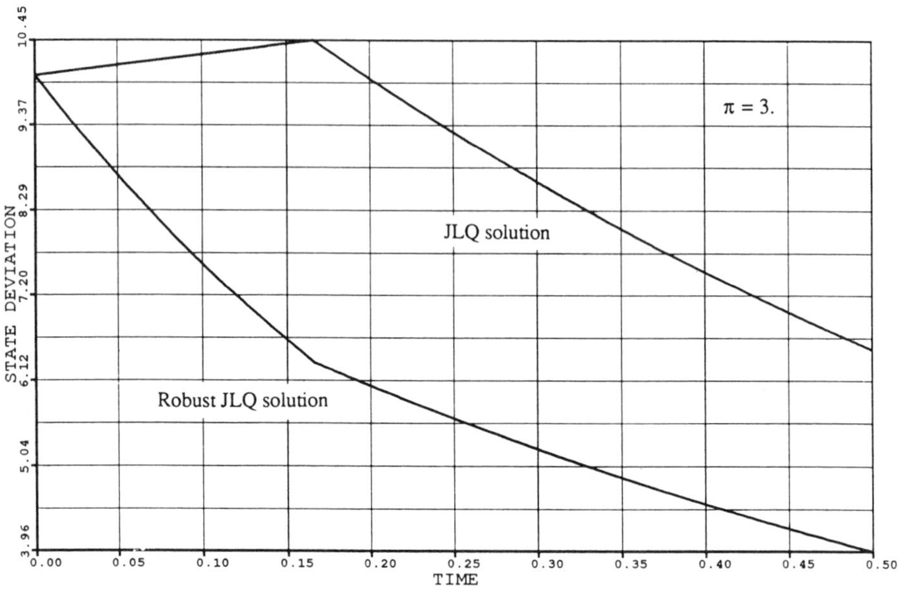

Figure 4.6 - JLQ / robust JLQ solutions.

Regardless of how long the system actually stays in regime 2, the robust solution guarantees that x_t exponentially decreases to equilibrium.

4.7 Notes and References

Various aspects of the robustness of a hybrid system with respect to sample path uncertainty have been studied. The results obtained cover both the analysis and synthesis viewpoints. To evaluate the robustness of a mean optimal solution, the designer can use either the bounds of §4.3 on the instantaneous cost or the simpler variance computation of §4.2. It may even be decided to compute the realized cost distribution. When this analysis indicates an unsatisfactory behavior, the alternative equalizing and minimax strategies of §4.5 may then be considered.

Most of the analysis tools presented in §4.2 and §4.3 were adapted from the work of Sworder and his students (Sworder and Choi, 1976ab, Kawauchi and Sworder, 1978, Kawauchi, 1977), including the example used to illustrate the variance computation. Though most stochastic control results are limited to mean performance optimization, some related research may be found in an LQG setting: (Sain, 1966) solved a "minimal variance with fixed mean" problem in an open-loop fashion with a 4n (n = dim x) two point boundary value problem as an intermediate step. Further work was later reported in (Sain and Liberty, 1971, Liberty and Hartwig, 1978). Using the exponential of a quadratic integral as the cost function, (Jacobson, 1973) showed that there exist some degrees of freedom to shape the distribution of the realized cost under the optimal control law. More precisely, a tuning parameter is exhibited which indirectly balances the weight given to high order and low order moments.

The key step for achieving more precise results for hybrid systems was the connection with the theory of statistical games explained in §4.5. The origin of this theory may be found in (Borel, 1921) but the book by (Wald, 1947) is certainly the first systematic presentation, including such long-lasting contributions as the sequential probability ratio test (SPRT) which is still at the heart of much current research in the detection literature (Basseville and Benveniste Eds, 1986). A suitable introduction to this rich field is constituted by the monographs (Blackwell and Girshick, 1954) and (Ferguson, 1967). Despite striking similarities between stochastic control and statistical game formulation, the tools of statistical decision-making appear to have made little headway in the domain of stochastic control systems, with notable exceptions such as (Sworder, 1965), (Pierce and Sworder, 1971) and their references.

The equalizing and minimax concepts of §4.5 are connected by a fundamental theorem of games theory which states that if an equalizing solution is also optimal in a bayesian sense then it is minimax (Ferguson, 1967). Due to the difficulty of a frontal attack on minimax optimization for general systems, this result is often used to obtain a minimax solution from an equalizing one. The problem is then reduced to the search for a distribution of Nature's decisions under which the equalizer is (extended) Bayes (Sworder, 1965). An exploration of this path in a related JLQ setting may be found in (Pierce and Sworder, 1971). Even though the analysis of §4.4 leads to a direct solution of the minimax problem, this indirect result provides a check-point. The optimal equalizer given in §4.5 by corollary 1 is clearly also minimax since Γ_{iw}^{**} is precisely the minimax gain (4.5.13). This can also be checked from the general theorem of (Ferguson, 1967) by exhibiting a distribution for Nature's choice with respect to which this equalizer is also Bayes. But this is trivially the case for the measure on the set of Nature choices assigning probability one to the event ω_0 defined as

$$r_t(\omega_0) = i_w \qquad t \in [t_0, t_f] \qquad (4.7.1)$$

(in other words the constant worst regime realization). For this measure the stochastic nature of the problem is degenerate (no jump) and $\Gamma_{i_w}^{**}$ is simply the LQ optimal gain corresponding to the i_wth regime. Being Bayes with respect to (4.7.1), the equalizer is also minimax, as we already knew from elementary arguments.

Regarding robustness with respect to parameter uncertainties, the JLQ regulator carries the robustness margins of its deterministic LQ counterpart. Considering a perturbation $B(r_t)u_t \rightarrow B(r_t)\, g(u_t, r_t)$ of the control input

$$\dot{x}_t = A_t x_t + B(r_t)\, g(u_t, r_t)$$

it can be shown that a stable (ESMQ) jump linear system remains stable provided $g(-, r_t)$ is contained, for every $r_t \in \mathbf{S}$, in a cone, as shown in figure 4.7.

Limiting ourselves to $g(u_t, i) = \alpha_i u_t$, the above admissible region is simply characterized by $\frac{1}{2} \leq \alpha_i \leq \infty$; in other words a 50% gain reduction tolerance and infinite gain margin. The proof of this result, using stochastic Liapunov functions, follows closely that of the LQ case presented in detail in the monograph (Safonov, 1980). It should be noted, however, that these margins, good as they may sound, do not provide a completely satisfactory result, in particular because they measure robustness under a state feedback law while in practice the state is rarely measured. Robustness analysis and robustness-directed synthesis therefore remain very active fields of research.

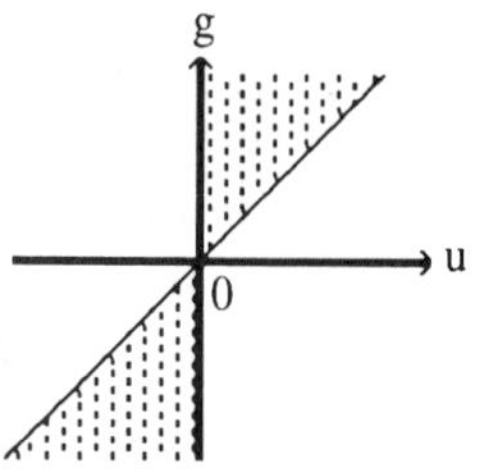

Figure 4.7 - Admissible non-linearities.

Theorem 4.4 provides a robust solution to the regulator problem when the regime is measured (switching gain feedback). In some situations, however, this measurement is not available, or the detection device estimating the regime may be malfunctioning, and a back-up solution is desirable to stabilize the system with a regime-independent law. Using a technique from (Petersen, 1987), such a solution was recently proposed in (Mariton, 1988) in terms of a nonlinear feedback control law: for each state value x, it is shown how to compute the control of minimum absolute value which makes the considered stochastic Liapunov function decrease.

References

Blackwell, D. and Girshick, M.A. (1954). Theory of Games and Statistical Decisions, J. Wiley, New York.

Borel, E. (1921). La théorie du jeu et les équations intégrales à noyau symétrique, C.R. Acad. Sci. Paris, 173: 1304.

Ferguson, T. (1967). Mathematical Statistics: a Decision Theoretic Approach, Academic, New York.

Garsia, A.M. (1973). Martingale Inequalities, Seminar Notes on Recent Progress, W.A. Benjamin Inc., Reading.

Jacobson, D.H. (1973). Optimal stochastic linear systems with exponential performance criteria and their relation to deterministic differential games, IEEE Trans. Aut. Control, AC-18: 124.

Kawauchi, B.H. (1977). Determination of the instantaneous cost functions for optimally controled stochastic systems, Ph.D. Dissertation, Univ. Southern California, Los Angeles.

Kawauchi, B.H. and Sworder, D.D. (1978). Instantaneous cost functions in optimally controlled stochastic systems, Proc. 17th IEEE Conf. Decision Control, San Diego, pp.484-489.

Liberty, S.R. and Hartwig, R.C. (1978). On the essential quadratic nature of LQG control performance measure cumulants, Inform. Control, 32: 276.

Mariton, M. (1988). Stabilization of systems with multiple regimes and Markov transitions, submitted for publication.

Morrison, J.A. (1972). Moments and correlation functions of solutions of some matrix differential equations, J. Math. Physics, 13: 299.

Petersen, I.R. (1987). A procedure for simultaneously stabilizing a collection of single input linear systems using non-linear state feedback, Automatica, 23: 33.

Pierce, B.D. and Sworder, D.D. (1971). Bayes and minimax controllers for a linear system with stochastic jump parameters, IEEE Trans. Aut. Control, AC-16: 300.

Sain, M.K. (1966). Control of linear systems according to the minimal variance criterion - a new approach to the disturbance problem, IEEE Trans. Aut. Control, AC-11: 118.

Sain, M.K. and Liberty, S.R. (1971). Performance-measure densities for a class of LQG control systems, IEEE Trans. Aut. Control, AC-16: 431.

Safonov, M.G. (1980). Stability and Robustness of Multivariable Feedback Systems, MIT Press, Cambridge.

Sworder, D.D. (1965). Minimax control of discrete time stochastic systems, SIAM J. Control Optim., A-2: 433.

Sworder, D.D. and Choi, L.L (1976a). Stationary cost densities for optimally controlled stochastic systems, IEEE Trans. Aut. Control, AC-21: 492.

Sworder, D.D. and Choi, L.L (1976b). Density functions for random matrix equations, Math. Progr. Study, 6: 246.

Sworder, D.D. (1977). A simplified algorithm for computing stationary cost variance for a class of optimally controled jump parameter systems, IEEE Trans. Aut. Control, AC-22: 236.

Wald, A. (1947). Sequential Analysis, J. Wiley, New York.

5

The Jump Linear Quadratic Gaussian Problem

5.1 Introduction

The hybrid models considered so far were described by piecewise deterministic differential equations: except for sudden regime transitions governed by a Markov chain, the state evolution obeys a deterministic equation. Similarly it was assumed that perfect observation of the plant state and regime is available. Starting with this chapter, more complex situations are studied where "noises" affect the plant behaviour, both in the state evolution and in the measurement channels.

This is motivated by recalling the typical applications of chapter 1. In the solar thermal receiver, an important measurement is the metal temperature channel which is implemented with a thermocouple sensor which delivers a value close to the true value plus a fluctuating corruption signal adequately modelled as "noise". Also in the target tracking application, the environment (wind gusts, etc.) affects both target and tracker trajectories in a random fashion, thus leading to "noise driven" state dynamics.

The notion of noise in technological systems was defined as early as the beginning of World War II, and the mathematical framework of stochastic differential equations is now available. Appendix 1 gives an overview of the required notions from stochastic integral calculus and introduces the mathematical counterpart of the "white noise" conveniently used in our engineering applications.

The influence of noise on hybrid systems is complex and, for clarity of exposition, we have decided to proceed gradually by assuming in this chapter that the regime r_t is still perfectly measured. Noises therefore affect only the state dynamics and observations, but the regime measurement channel remains uncorrupted. The more general situation is postponed until chapters 6 and 7 where control in the presence of regime uncertainties is studied.

The preliminary results of this chapter are also valuable in the situation where the regime is not measured: it will be seen in chapter 7 that the control design in such a case becomes very involved. A reasonable approximation is thus to use an estimate of the regime cascaded with a solution designed for the exact regime measure situation. In other words, the solutions of the present chapter can be used in conjunction with the regime estimates of chapter 6 in a "Certainty Equivalence Principle" structure (see appendix 2 and §5.3 below).

In §5.2 an even simpler situation is considered when both the state and regime are measured for a dynamic equation with additive and multiplicative noises. This allows a quantitative analysis of the influence of uncertainties on the JLQ regulator and it constitutes a complement to the parameter robustness studies of chapter 4: a poor knowledge of, say, the dynamic matrices A_i, i = 1 to N, can be modelled by a state multiplicative noise. The corresponding regulator is obtained and, because gaussian noises are assumed, the name Jump Linear Quadratic Gaussian (JLQG) regulator is proposed.

When noises corrupt the plant variables, direct feedback of the available signals results in high actuator activity and undesired high frequency excitation of the plant. To avoid these effects, the classical approach is to low-pass the corrupted signal to filter out as much noise as possible. Clearly the bandwidth must be large enough to allow transmission of the underlying useful information, thus leading to another occurrence of the well-known sensitivity/accuracy trade-off. Concepts of stochastic processes theory and, in particular, optimal filters can in fact be viewed as a way of making this trade-off automatically, at least for those applications where reliable models are available.

For ξ_t a noise corrupted plant variable and O_t the σ-algebra generated by the plant observations up to time t, the best low-pass operation is provided by the conditional expectation $E\{\xi_t \mid O_t\}$. It turns out that the adjective "best" in fact indicates optimality with respect to the a posteriori mean square error (see appendix 1). In this chapter the variable ξ_t of interest is the plant state, and §5.3 shows that the optimal filter, under the assumption r_t measured, has a structure very close to that of the usual Kalman filter of the LQG problem.

Finally, the influence of impulsive disturbances on the state trajectories is discussed in §5.4 by means of a Poisson noise model.

5.2 The JLQG Regulator

The linear hybrid model is generalized to include the presence of additive and multiplicative noises

$$dx_t = A(r_t)x_t dt + B(r_t)u_t dt + D(x_t, r_t)dw_{1t} + E(u_t, r_t)dw_{2t} + F(r_t)dw_{3t} \quad (5.2.1)$$

The noises w_{it}, $i = 1,3$, are independent Wiener processes on $\mathbf{R}^{n_i}$ and the interpretation of this diffusion with markovian switching coefficients is as a collection of piecewise defined Ito stochastic differential equations (see appendix 1 and its references for an introduction to stochastic integrals). The definition of (5.2.1) must be completed by a description of what happens at jump times when the dynamics go from one well-defined stochastic differential to another. In this section it is considered that x_t is continuous at these instants, but §5.4 below extends the results to systems where x_t experiences a discontinuity of random magnitude when the regime changes.

Matrices D, E and F, of dimensions respectively $n \times n_1$, $n \times n_2$ and $n \times n_3$, are regime-dependent so that the noises influence the plant with an intensity which is modulated by the value of r_t. Misalignment angles of satellite main thrusters are affected by failures of the swivel mechanism and the uncertainty surrounding a thruster firing is adequately described by a control-dependent noise with a larger intensity in failed regimes. Another example is provided by the target tracking application where the disturbance acceleration in the target kinematics model depends on the target attitude. This can be included in (5.2.1) by varying accordingly the additive noise input matrix F as a function of the target orientation captured by r_t. When failures affect the plant dynamics there may also be regimes where there is great uncertainty on the expected value of A; this is easily quantified by the state-dependent matrix $D(x_t,r_t)$.

The model (5.2.1) is, however, too general, especially because of the state (D) and control (E) nonlinearities, and attention is immediately restricted to the case where D and E are in fact linear in x and u.

$$D(x_t, r_t) = \sum_{k=1}^{n} x_{kt} D_k(r_t)$$

$$E(u_t, r_t) = \sum_{k=1}^{m} u_{kt} E_k(r_t)$$

where x_{kt} and u_{kt} are the k-th component of, respectively, the state x_t and control u_t at time t. Associated with this special case, the endomorphisms **D** and **E** are defined for P a symmetric n x n matrix

$$\mathbf{d}_{k\ell}(r_t, P) = \mathrm{trace}(D_k(r_t)'P\, D_\ell(r_t)) \qquad \mathbf{D} = (\mathbf{d}_{k\ell})_{k,\ell = 1,n}$$

$$\mathbf{e}_{k\ell}(r_t, P) = \mathrm{trace}(E_k(r_t)'P\, E_\ell(r_t)) \qquad \mathbf{E} = (\mathbf{e}_{k\ell})_{k,\ell = 1,m}$$

It then appears that w_1 acts as a state multiplicative noise and w_2 as a control multiplicative noise. As usual the regime dependence of **D** and **E** is often denoted by an index.

With this model, the Jump Linear Quadratic Gaussian is defined by considering again the averaged quadratic cost of chapter 3

$$J = E \{ \int_{t_0}^{t_f} (x_t'Q_1x_t + u_t'Q_2u_t)\, dt \mid x_0, i_0, t_0 \}$$

For the present time it is still assumed that both the state x_t and regime r_t of the hybrid model are measured. Admissible control laws are restricted to instantaneous feedback laws $u_t = U(x_t, r_t, t)$ with once again the smoothness conditions of §3.2 on U (see also appendix 2), but, due to the Markov property of the pair (x_t, r_t) which is preserved by (5.2.1), this encompasses the class of $X_t \vee R_t$ measurable solutions where X_t and R_t are the state and regime observations σ-algebras.

Denoting by $\mathscr{Z}$ the class of admissible solutions, the JLQG problem is then stated formally as

$$\underset{\mathscr{U}}{\text{Min}}\ J \quad \text{subject to (5.2.1)} \tag{5.2.2}$$

The dynamic programming approach is adopted so that the first step is to compute the generator of the pair (x_t, r_t) for (5.2.1) and the regime Markov chain. The result is given by the following theorem.

Theorem 5.1

For $g(x, r)$ a function from $\mathbf{R}^n \times \mathbf{S}$ into $\mathbf{R}$ such that g and g_x are continuous in x for any r and such that $| g(x, r) | < \alpha (1 + | x |)$ for a constant α, the generator $\mathscr{L}_u$ of (x_t, r_t) under an admissible control action u, for x_t solution of (5.2.1) and r_t a Markov chain on $\mathbf{S}$ with transition rates matrix Π, is given by

$$\mathscr{L}_u\, g(x,i) = (A_i x + B_i u)' g_x(x,i) + \sum_{j=1}^{N} \pi_{ij}\, g(x,j)$$

$$+ \frac{1}{2} \text{trace}[D(x,i)' g_{xx}(x,i) D(x,i) + E(u,i)' g_{xx}(x,i)\, E(u,i) + F_i' g_{xx}(x,i) F_i] \tag{5.2.3}$$

Proof:

From the definition of the generator (appendix 1) the quantity to be evaluated is

$$E\,\{g(x_t, r_t) \mid x_{t-} = x, r_{t-} = i\,\}$$

Observe that

$$g(x_t, r_t) = g(x, i) + dx_t' g_x(x,i) + \frac{1}{2}\, dx_t' g_{xx}(x,i)\, dx_t + \sum_{j=1}^{N} g(x,j) d\phi_{tj}$$

where it is recalled that ϕ_t is the regime indicator with components $\phi_{ti} = 1$ if $r_t = i$, 0 otherwise. Taking the conditional expectation for independent Wiener processes w_{1t}, w_{2t} and w_{3t} then gives

$$g(x,i) + (A_ix + B_iu)'g_x(x,i)\,dt + \frac{1}{2}\,\text{trace}[D(x,i)'g_{xx}(x,i)D(x,i)]\,dt$$

$$+ \frac{1}{2}\,\text{trace}[E(x,i)'g_{xx}(x,i)E(x,i)]\,dt \;+ \frac{1}{2}\,\text{trace}[F_i'g_{xx}(x,i)F_i]\,dt$$

$$+ \sum_{j=1}^{N} \pi_{ij}\; g(x,j)\,dt$$

where the last term is deduced from the martingale representation of ϕ_t. Finally, the limit of $1/dt\,(E\,\{g(x_{t+dt}, r_{t+dt}) \mid x_t = x, r_t = i\} - g(x,i))$ as dt goes to zero produces (5.2.3) of the theorem. □

This result does not use the linearity of D and E in x and u and indeed it is valid for more general kinds of noise modulation.

The multiplicative and additive noises manifest themselves in the generator through the quadratic variation term $dx_t'dx_t$ in dg. This term was neglected in the piecewise deterministic case of chapter 3 because it was then of order dt^2, but, due to the dw_{1t}, dw_{2t} and dw_{3t} presence in dx_t, its contribution is now of order dt and it may not be deleted.

However, when $g(x_t, r_t)$ is quadratic in x_t and D and E linear in x and u, equation (5.2.3) can be simplified using the endomorphisms **D** and **E** previously introduced. This is why it is so interesting to consider quadratic cost functionals, as is illustrated by the optimal JLQG regulator computation.

Theorem 5.2

The solution of the optimization problem (5.2.2) is

$$u_t^* = - [Q_{2i} + \mathbf{E}_i(\Lambda_i)]^{-1}\, B_i'\Lambda_i x_t \quad \text{when } r_t = i \tag{5.2.4}$$

with the Λ_is, i = 1 to N, obtained from

$$-\dot{\Lambda}_i = A_i'\Lambda_i + \Lambda_i A_i - \Lambda_i B_i [Q_{2i} + \mathbf{E}_i(\Lambda_i)]^{-1} B_i'\Lambda_i + Q_{1i} + \mathbf{D}_i(\Lambda_i) + \sum_{j=1}^{N} \pi_{ij} \Lambda_j$$

$$\Lambda_i(t_f) = 0 \tag{5.2.5}$$

The minimal cost is

$$J^* = x_0'\Lambda_{i_0}(t_0)x_0 + \mu_{i_0}(t_0)$$

for $x_{t_0} = x_0$ and $r_{t_0} = i_0$, where the μ_is, i = 1 to N, satisfy

$$-\dot{\mu}_i = \text{trace}\,(F_i'\Lambda_i F_i) + \sum_{j=1}^{N} \pi_{ij}\,\mu_j$$

$$\mu_i(t_f) = 0 \tag{5.2.6}$$

<u>Proof:</u>

Introducing the conditional cost-to-go V(x, i, t) as

$$V(x, i, t) = \min_{\mathscr{U}} E\{ \int_t^{t_f} (x_s'Q_1x_s + u_s'Q_2u_s)\, ds \mid x_t = x, r_t = i, t \}$$

the verification theorem of dynamic programming leads to Bellman's equation

$$0 = \min_{u \in \mathbf{R}^m} \{V_t(x,i,t) + \mathscr{L}_u V(x,i,t) + x_t'\,Q_1x_t + u_t'Q_2u_t\}$$

$$\text{for } t_0 \le t \le t_f, \quad (x,i) \in \mathbf{R}^n \times \mathbf{S}$$

and the boundary condition $V(x,i,t_f) = 0$. The generator $\mathscr{L}_u$ is given by theorem 5.1 and the minimization can be performed explicitly when E is linear in u

$$u_t^* = -\frac{1}{2}\,[Q_{2i} + \frac{1}{2}\,\mathbf{E}_i\,(V_{xx})]^{-1}\, B_i'V_x(x,i,t) \text{ when } r_t = i.$$

With this expression of the optimal control, Bellman's equation becomes a second order partial differential equation for V which is conveniently solved by looking for a quadratic form solution

$$V(x,i,t) = x_t' \Lambda_i x_t + \mu_i$$

where the time dependence of the Λ_is and μ_is is not explicited. Using such a V, expression (5.2.4) for u_t^* is immediately obtained. Next, bringing this back into Bellman's equation we obtain, at the minimum

$$0 = x_t'\dot{\Lambda}_i x_t + \dot{\mu}_i + 2(A_i x_t - B_i[Q_{2i} + \mathbf{E}(\Lambda_i)]^{-1}B_i'\Lambda_i x)'\Lambda_i x_t + \sum_{j=1}^{N} \pi_{ij} (x_t'\Lambda_j x_t + \mu_j)$$

$$+ x_t'\mathbf{D}_i(\Lambda_i)x_t + x_t'\Lambda_i B_i[Q_{2i} + \mathbf{E}(\Lambda_i)]^{-1}\mathbf{E}(\Lambda_i)[Q_{2i} + \mathbf{E}(\Lambda_i)]^{-1}B_i'\Lambda_i x_t$$

$$+ \text{trace } (F_i'\Lambda_i F_i)$$

$$+ x_t'Q_{1i} x_t + x_t'\Lambda_i B_i[Q_{2i} + \mathbf{E}(\Lambda_i)]^{-1} Q_{2i}[Q_{2i} + \mathbf{E}(\Lambda_i)]^{-1}B_i'\Lambda_i x_t$$

The right-hand side must be zero for any (x,i,t) so that the only solution is that the quadratic and constant coefficients are zero. Grouping terms, this eventually gives (5.2.5) and (5.2.6) of the theorem for, respectively, the quadratic and constant terms. Finally, the expression of the optimal cost results from the definition of the cost-to-go for $t = t_0$. □

The regulator of theorem 5.2 contains as special cases several well-known regulators. When the system is deterministic ($\Pi = 0$, $\mathbf{D}_i = 0$, $\mathbf{E}_i = 0$, $F_i = 0$, i = 1 to N), we recover the usual LQ regulator, while if only the addditive and multiplicative noises disappear ($\Pi \neq 0$) the JLQ regulator of chapter 3 is again obtained. The LQG situation (Wonham, 1971) is finally recovered by letting $\Pi = 0$ for non-zero $\mathbf{D}_i$, $\mathbf{E}_i$ and F_i. This imbrication of results is captured by the name JLQG regulator given to the solution of theorem 5.2.

The influence of jumps manifests itself through the coupling terms $\sum_j \pi_{ij} \Lambda_j$ and $\sum_j \pi_{ij} \mu_j$ in (5.2.5) and (5.2.6). It forces the integration of the N equations (5.2.5) to obtain the optimal feedback gains $\Gamma_i^* = - [Q_{2i} + \mathbf{E}_i(\Lambda_i)]^{-1}B_i'\Lambda_i$, i = 1 to N. The μ_i equations are a set of N linear scalar differential equations for known Λ_is, and this is quite easily integrated, backwards in time, from $\mu_i(t_f) = 0$, i = 1 to N. However, the value of μ_i does not affect the control law but only the minimal cost J*, and often its computation is neglected. As in the LQG setting it is found that, when the state and regime are both exactly measured, the additive state noise does not influence the optimal regulator; of course, this is no longer the case when noisy observations are considered. Regarding the N-coupled matrix equations (5.2.5) for the Λ_is, they are more difficult to solve. A peculiarity is the $\mathbf{E}_i(\Lambda_i)$ term which makes (5.2.5) rational, and not quadratic, in Λ_i. Equations (5.2.5) are therefore not Riccati-type equations, even though, due to their familiar structure, they are often called "Riccati-like" equations. Algorithms should be adapted from the numerical integration schemes studied in (Incertis, 1982) for the jumpless case ($\Pi = 0$).

Looking at (5.2.4) and (5.2.5), it is possible qualitatively to describe the influence of the multiplicative noises on the optimal regulator. In particular the following phenomenon is observed: the larger the control multiplicative noise (E_i large) the smaller the gain; on the other hand a larger state multiplicative noise (D_i large) gives a larger gain. The interpretation is that when the control action is very uncertain (E_i large) the controller should behave cautiously and reduce its gain while the destabilizing influence of a large state multiplicative noise should be countered by a larger gain.

As in the JLQ case, it seems possible to simplify the computation of the JLQG regulator by restricting attention to the asymptotic ($t_f - t_0 \to \infty$) problem. Since the additive noise does not influence the regulator gains, they are now deleted (this avoids technicalities about the definition of (5.2.1) under a persistently exciting wide-band signal). From chapter 3 it is known that the asymptotic behaviour of the JLQG regulator should be discussed in terms of the conditions of stabilizability and detectability. Using the stabilizability definition concerning mean square stability (ESMS, see chapter 2) the following conditions are obtained.

<u>Theorem 5.3</u>

The noisy jump linear system (5.2.1) with transition rates matrix Π is stochastically stabilizable if there exists feedback gains Γ_i, i = 1, N, such that for some positive definite matrices Q_i, i = 1, N, the solutions P_i, i = 1, N, of the set of coupled equations

$$\tilde{A}_i'P_i + P_i\tilde{A}_i + \Gamma_i'\mathbf{E}_i(P_i)\Gamma_i + \mathbf{D}_i(P_i) + Q_i + \sum_{j=1}^{N} \pi_{ij} P_j = 0 \tag{5.2.7}$$

are positive definite (where $\tilde{A}_i = A_i + B_i\Gamma_i$, i = 1 to N).

Proof:

The quadratic function

$$V(x,i,t) = x'P_ix$$

satisfies the conditions to be a possible Liapunov function (see chapter 2, §2.4). With the above generator we compute

$$\mathcal{L}V = x'(\tilde{A}_i'P_i + P_i\tilde{A}_i + \Gamma_i'\mathbf{E}_i(P_i)\Gamma_i + \mathbf{D}_i(P_i) + Q_i + \sum_{j=1}^{N} \pi_{ij} P_j)x$$

so that, for the P_is solutions of (5.2.7)

$$\mathcal{L}V = -x'Q_i x$$

and theorem 2.3 then guarantees that the closed-loop system is ESMS. □

Equations (5.2.7) play the same role in relation to (5.2.5) that the Liapunov equations played in relation to the Riccati equations of the JLQ regulator, and are therefore referred to as "Liapunov-like" equations. Considering observations with additive and multiplicative noises, a dual notion of detectability is immediately obtained as in chapter 2 (see §2.5) and the following characterization of the asymptotic JLQG regulator is proposed.

Theorem 5.4

The limits $\bar{\Lambda}_i = \lim_{t_0 \to -\infty} \Lambda_i$, i = 1, N, exist as the unique positive semi-definite solutions of the set of coupled algebraic equations

$$0 = A_i'\bar{\Lambda}_i + \bar{\Lambda}_iA_i - \bar{\Lambda}_iB_i[Q_{2i} + \mathbf{E}(\bar{\Lambda}_i)]^{-1}B_i'\bar{\Lambda}_i + Q_{1i} + \mathbf{D}(\bar{\Lambda}_i) + \sum_{j=1}^{N} \pi_{ij}\bar{\Lambda}_j$$

and the closed-loop system is exponentially stable in mean square if and only if the system is stochastically stabilizable and detectable.

Proof:

It imitates the proof of theorem 3.3 in chapter 3 and is therefore omitted. □

In the LQG setting it has been demonstrated that the convergence of the asymptotic regulator is ruled by the (multiplicative) noise intensities and that there exist intensity thresholds above which convergence does not occur. The name Uncertainty Threshold Principle (UTP) was coined in (Athans et al., 1977) to describe these phenomena. It has recently been noted that the UTP is in fact a property of the Ito integral representation of the noisy process dynamics and that it disappears if the Stratonovitch correction term is appropraitely included (Bernstein and Hyland, 1985).

Despite this cautionary remark, it is interesting to study what kind of UTP-like phenomena may occur in the JLQG situation with (5.2.1) an Ito stochastic differential equation.

First the findings of (Athans et al., 1977) in the LQG case are recalled:

- When only control noise is present ($D_i = 0$, $i = 1$ to N), divergence may occur only if the open-loop system is unstable.

-When only state noise is present ($E_i = 0$, $i = 1$ to N), divergence may occur at large noise intensities, the interpretation being a noise-induced plant destabilization.

Of course, as the JLQG case contains the LQG case, the two qualitative remarks above also hold. But new phenomena linked to the regime transitions are observed, as illustrated by the following example.

Example:

Consider a scalar system with two modes and a matrix of transition rates $\Pi = \begin{bmatrix} -\pi_1 & \pi_1 \\ \pi_2 & -\pi_2 \end{bmatrix}$. Replacing all upper-case letters by their lower-case counterparts for this scalar example, equation (5.2.5) becomes

$$- \dot{\lambda}_1 = 2a_1\lambda_1 - \lambda_1{}^2 b_1{}^2[q_{21} + \lambda_1 e_1{}^2]^{-1} + q_{11} + \lambda_1 d_1{}^2 - \pi_1\lambda_1 + \pi_1\lambda_2$$
$$- \dot{\lambda}_2 = 2a_2\lambda_2 - \lambda_2{}^2 b_2{}^2[q_{22} + \lambda_2 e_2{}^2]^{-1} + q_{12} + \lambda_2 d_2{}^2 - \pi_2\lambda_2 + \pi_2\lambda_1$$

If it occurs, divergence leads to large values of λ_1 and λ_2 so that, approximately,

$$- \dot{\lambda}_1 \approx 2a_1\lambda_1 - \lambda_1 b_1{}^2/e_1{}^2 + \lambda_1 d_1{}^2 - \pi_1\lambda_1 + \pi_1\lambda_2$$
$$- \dot{\lambda}_2 \approx 2a_2\lambda_2 - \lambda_2{}^2 b_2{}^2/e_2{}^2 + \lambda_2 d_2{}^2 - \pi_2\lambda_2 + \pi_2\lambda_1$$

which can be written in matrix form

$$\dot{\begin{bmatrix} \lambda_1 \\ \lambda_2 \end{bmatrix}} = \begin{bmatrix} m_1+\pi_1 & -\pi_1 \\ -\pi_2 & m_2+\pi_2 \end{bmatrix} \begin{bmatrix} \lambda_1 \\ \lambda_2 \end{bmatrix} \quad \text{with } m_i = -2a_i - d_i^2 + b_i^2/e_i^2 \quad , i=1,2.$$

The quantities m_1 and m_2 are related to the jumpless LQG situation where the λ_1 and λ_2 equations uncouple ($\pi_1 = \pi_2 = 0$). They are precisely the quantities analyzed in (Athans et al., 1977) to obtain the above remarks. Jumps result in a coupling described by the matrix

$$M = \begin{bmatrix} m_1+\pi_1 & -\pi_1 \\ -\pi_2 & m_2+\pi_2 \end{bmatrix}$$

Recalling that this vector differential equation is integrated backwards in time, it converges if and only if the eigenvalues of M have positive real parts. The characteristic polynomial is

$$z^2 - z(m_1 + \pi_1 + m_2 + \pi_2) + (m_1 + \pi_1)(m_2 + \pi_2) - \pi_1\pi_2 = 0$$

It has two real zeros with sum $S = m_1 + \pi_1 + m_2 + \pi_2$ and product $P = (m_1 + \pi_1)(m_2 + \pi_2) - \pi_1\pi_2$. The convergence condition is therefore (two positive zeros)

$$(m_1 + \pi_1)(m_2 + \pi_2) \geq \pi_1\pi_2 \quad \text{and} \quad m_1 + \pi_1 + m_2 + \pi_2 \geq 0$$

From this analysis, the following UTP-like properties can be deduced

- For a noiseless controlled system ($d_i = e_i = 0$, $b_i \neq 0$, $i = 1, 2$) convergence occurs unconditionally since $m_1 = m_2 = +\infty$. This simply recovers the usual LQ (and JLQ, see chapter 3) result.

- Role of the regime average life-time: assuming regime 2 is absorbing ($\pi_2 = 0$) with $m_2 = 0$ the convergence condition reduces to $m_1 + \pi_1 \geq 0$. This means that the JLQG regulator can still converge with $m_1 < 0$ (its LQG counterpart then diverges) provided π_1 is large enough. As observed in chapter 2 this is an averaged dynamics property where a short average life-time ($1/\pi_1$ small) compensates for the detrimental influence of the regime ($m_1 < 0$). The threshold value of π_1 is

$$\pi_{1T} = 2a_1 + d_1^2 - b_1^2/e_1^2$$

and convergence occurs for any $\pi_1 \geq \pi_{1T}$. The expression of π_{1T} illustrates the role of multiplicative noises: the larger d_1, the larger is π_{1T}, the larger e_1, the smaller is π_{1T}. A large state multiplicative noise destabilizes the plant and the life-time has to be shorter; the control multiplicative noise has a symmetric role.

- Couplings through jumps: for $\pi_1 = \pi_2 = \pi$ large we have $S \approx 2\pi$ and $P \approx \pi(m_1 + m_2)$ so that the convergence conditions reduce to $m_1 + m_2 \geq 0$. Again there is an averaging between communicating regimes and convergence may occur with $m_1 \leq 0$ provided m_2 is positive enough (see the related discussion of chapter 2 on the stability of state trajectories).

5.3 A Kalman Filter

The JLQG regulator of §5.2 captures the influence of noises on hybrid systems, but it is still based on the assumption that perfect, noise-free measurement channels are available for the state and regime variables. A more realistic situation is when noises also affect the observations, and the simplest case is then noisy state measurements with a known regime

$$dy_t = C(r_t)\, x_t\, dt + H(r_t)\, dv_t \tag{5.3.1}$$

where C is a regime-dependent p x n matrix and v_t a Wiener process on $\mathbf{R}^{n_4}$. The p x n_4 matrices H_i, i = 1 to N, modulate the noise intensity and are supposed to satisfy the non-degeneracy condition H_iH_i' invertible, i = 1 to N. This condition states that noise corrupts all the observations. The model (5.3.1) produces a noisy measurement of x_t which in turn obeys an equation like (5.2.1) without multiplicative noise

$$dx_t = A(r_t)\, x_t\, dt + F(r_t)\, dw_t \tag{5.3.2}$$

Dropping multiplicative noises is interesting because it simplifies the problem but the more general case with (5.2.1) observed through (5.3.1) is considered again in §5.5.

As explained in the introduction, the way to extract information on x_t from the knowledge of y_t and r_t is to form the a posteriori expectation

$$\hat{x}_t = E\{x_t \mid Y_t \vee R_t\}$$

where Y_t and R_t are, respectively, the y_t and r_t observation σ-algebras. Due to the Markov property this is equivalent to

$$\hat{x}_t = E\{x_t \mid Y_t, r_t\} \tag{5.3.3}$$

Indeed it would be desirable to go one step further and to compute $\hat{x}_t$ recursively, that is, in a way where new measurements (dy_t, r_t) are used as they become available to refine the current estimate with as little additional computation as possible. Such a recusive form is not explicit in (5.3.3) where it seems that to go from t to t + dt requires a complete evaluation of the expectation conditioned on Y_{t+dt}. Intuitively $\hat{x}_t$ already extracts the information in Y_t and it should be possible to compute $\hat{x}_{t+dt}$ from $\hat{x}_t$ and, furthermore, the modification $(\hat{x}_{t+dt} - \hat{x}_t)$ should depend only on $\hat{x}_t$ and the new piece of information dy_t. This intuitive reasoning is derived from the well-known Kalman filter philosophy and also turns out to be valid in the presence of (observed) regime transitions. The key tool in this respect is the optimal filter representation theorem (see appendix 1) from which it is possible to derive $\hat{x}_t$ as the solution of a stochastic differential equation driven by the observations y_t and r_t.

Assuming that the initial state x_{t_0} is a centered random variable with $E\{x_{t_0} x_{t_0}'\} = X_0$ and that the state w_t and observation v_t noises are independent, the result is given as

Theorem 5.5

The optimal state estimate $\hat{x}_t$

$$d\hat{x}_t = A_i \hat{x}_t dt + X_i C_i'(H_iH_i')^{-1} (dy_t - C_i \hat{x}_t dt) \text{ when } r_t = i \tag{5.3.4}$$

with $\hat{x}_{t_0} = 0$ where the X_is are solutions of a set of N coupled Riccati equations

$$\dot{X}_i = A_iX_i + X_i A_i' - X_i C_i'(H_iH_i')^{-1} C_iX_i + F_i F_i' + \sum_{j=1}^{N} \pi_{ij} x_j$$

$$X_{it_0} = X_0 \phi_{it_0} \tag{5.3.5}$$

The minimal error variance is trace X_i when $r_t = i$.

Proof:

From the filter representation theorem (see appendix 1), the computation of the optimal estimate is decomposed into an extrapolation of the model and an update on the innovation martingale. In the present situation this can be written

$$E\{A(r_t)x_t \mid Y_t \vee R_t\} = A_i \hat{x}_t$$
$$\text{when } r_t = i$$
$$E\{C(r_t)x_t \mid Y_t \vee R_t\} = C_i \hat{x}_t$$

so that, when $r_t = i$, $d\hat{x}_t$ is given by

$$d\hat{x}_t = A_i \hat{x}_t dt + K(H_iH_i')^{-1} (dy_t - C_i \hat{x}_t dt)$$

where the H_iH_i' term normalizes the noise intensity. The gain expression is

$$K = E\{(x_t - \hat{x}_t) (C(r_t)x_t - C(r_t)x_t)' \mid Y_t \vee R_t\}$$

Observing the regime, it is natural to introduce

$$K_i = E\{x_tx_t' - \hat{x}_t\hat{x}_t' \mid Y_t, r_t = i\} C_i'$$

The remaining step is therefore to compute X_i defined as

$$X_i = E\{x_t x_t' - \hat{x}_t \hat{x}_t' \mid Y_t, r_t = i\}$$

so that filter can be implemented as

$$d\hat{x}_t = A_i \hat{x}_t\, dt + X_i C_i'(H_i H_i')^{-1} (dy_t - C_i \hat{x}_t\, dt) \quad \text{when } r_t = i$$

as in (5.3.4) of the theorem.

First observe that

$$d(x_t x_t' \phi_{ti}) = A_i x_t x_t' \phi_{ti}\, dt + x_t x_t' A_i' \phi_{ti}\, dt + d[x_t, x_t \phi_{ti}] + x_t x_t' d\phi_{ti}$$

but ϕ_{ti} being zero or one

$$d[x_t, x_t \phi_{ti}] = d[x_t \phi_{ti}, x_t \phi_{ti}] = F_i F_i'\, dt$$

On the other hand

$$d(\hat{x}_t \hat{x}_t' \phi_{ti}) = A_i \hat{x}_t \hat{x}_t' \phi_{ti}\, dt + \hat{x}_t \hat{x}_t' A_i' \phi_{ti}\, dt + d[\hat{x}_t, \hat{x}_t \phi_{ti}] + \hat{x}_t \hat{x}_t' d\phi_{ti}$$

with

$$d[\hat{x}_t, \hat{x}_t \phi_{ti}] = d[\hat{x}_t \phi_{ti}, \hat{x}_t \phi_{ti}] = X_i C_i'(H_i H_i')^{-1} C_i X_i\, dt$$

Applying again the optimal filter representation theorem to the second moment $x_t x_t' - \hat{x}_t \hat{x}_t'$, we then obtain

$$dX_i = A_i X_i\, dt + X_i A_i' dt + F_i F_i'\, dt + X_i C_i'(H_i H_i')^{-1} C_i X_i\, dt + \sum_{j=1}^{N} \pi_{ij} X_j\, dt \tag{5.3.6}$$

$$+ E\{(x_t - \hat{x}_t)(x_t - \hat{x}_t)'(C(r_t)x_t - C(r_t)x_t)' \mid Y_t \vee R_t\}(H_i H_i')^{-1}(dy_t - C_i \hat{x}_t dt)$$

where the $\sum_j \pi_{ij} X_j$ term comes from $d\phi_{ti}$. The $dy_t - C_i \hat{x}_t dt$ coefficient in the above expression is proportional to the conditional third order moment and vanishes because the initially gaussian x_{t_0} is propagated through linear equations with known coefficients (r_t

measured). The observation driving term being thus cancelled, (5.3.6) in fact reduces to an ordinary differential equation

$$\dot{X}_i = A_i X_i + X_i A_i' + F_i F_i' + X_i C_i'(H_i H_i')^{-1} C_i X_i + \sum_{j=1}^{N} \pi_{ij} X_j \, dt$$

with initial condition $X_{it_0} = X_0 \phi_{it_0}$ for a known r_{t_0}.

Finally the minimal error covariance is trace X_i when $r_t = i$ by the very definition of the X_is, i = 1 to N, which completes the proof. □

The time-varying filter gains $X_i C_i'(H_i H_i')^{-1}$ are computed from the N coupled Riccati equations (5.3.5). Algorithms similar to those proposed in chapter 3 can be used to integrate (5.3.5), the important point being that this is an off-line task while on-line the task is only to modulate the innovation term by the precomputed gains. It should also be stressed that the equations for the X_is, i = 1 to N, are ordinary differential equations. Looking back at (5.3.4) it is seen that this is crucially dependent on the gaussianness of the estimation error $x - \hat{x}$: if the third moment of $x - \hat{x}$ is not zero then the last term in (5.3.6) does not vanish and (5.3.5) is replaced by a stochastic differential equation driven by the innovations. As explained above, it is the assumption of regime availability that preserves the gaussianness of the estimation error, and when it is removed in chapter 6 some type of modified "innovation-driven Riccati equations" will have to be considered to compute the gains. It is clear that this will be a quite different class of filters with a large augmentation of on-line computations, since the X_is, i = 1 to N, can then no longer be pre-computed.

With the transformations

$$A_i \rightarrow A_i'$$
$$B_i \rightarrow C_i'$$
$$Q_{1i}^{1/2} \rightarrow F_i$$
$$Q_{2i}^{1/2} \rightarrow H_i$$

there is an obvious duality between the coupled Riccati equations for the Kalman filter and those of the JLQ regulator in chapter 3. It may be observed that couplings in $\sum_{j=1}^{N} \pi_{ij} \Lambda_j$

for the regulator become couplings in $\sum_{j=1}^{N} \pi_{ji} X_j$ for the filter, indicating that the matrix Π is transformed in Π' by duality. Of course this is simply a consequence of forward propagation for the X_is as opposed to backward propagation for the Λ_is. Thanks to this formal remark, the asymptotic behaviour of the Kalman filter can be treated by dualizing the analysis of the asymptotic JLQ regulator of chapter 3.

Theorem 5.6

As $t_0 \to -\infty$, the solutions X_i, i = 1, N, of (5.3.5) converge to the unique positive semi- definite solutions $\bar{X}_i$ of the set of coupled algebraic Riccati equations

$$0 = A_i\bar{X}_i + \bar{X}_i A_i' + F_i F_i' - \bar{X}_i C_i'(H_iH_i')^{-1} C_i\bar{X}_i + \sum_{j=1}^{N} \pi_{ji} \bar{X}_j \tag{5.3.7}$$

and the dynamics

$$\dot{\hat{x}}_t = (A(r_t) - X(r_t) C(r_t)'[H(r_t)'H(r_t)]^{-1} C(r_t)) \hat{x}_t$$

are stable (ESMQ) iff the system $[A_i', C_i', F_i', i = 1, N, \Pi]$ is stochastically stabilizable and detectable.

Proof:

Repeat the derivation of theorem 3.3 in chapter 3. □

Just as for the asymptotic JLQ regulator, theorem 5.6 is an important practical result since it provides a reason for replacing the filter with time-varying gains by

$$d\hat{x}_t = A_i \hat{x}_t dt - \bar{X}_iC_i'(H_iH_i')^{-1} (dy_t - C_i \hat{x}_t dt) \quad \text{when } r_t = i$$

where the gains $\bar{X}_i$, i = 1, N, can be pre-computed as the constant solutions of (5.3.7). Savings in terms of storage memory requirements are then obvious.

In theorem 5.5 optimality in the sense of minimal a posteriori quadratic error is achieved by computing the filter gains via the solution of a set of coupled Riccati equations. However, the obtained optimality might be discussed from an engineering point of view: while there is no doubt that optimization in general provides a powerful mathematical formulation for design problems, the minimality of a cost or performance measure is not per se sufficient information for the engineer. There are several reasons for this, some of which were discussed in chapter 4. First optimality is strictly contingent upon a set of hypotheses (gaussianness, exact knowledge of the A_is, F_is, etc.) which are idealizations of conditions practically encountered. Second, there are often "hidden" objectives apart from the optimized index. Most prominent is the performance/complexity trade-off which would lead to rejection of solutions providing marginal cost improvement at the expense of increased hardware/software requirements.

With respect to this last consideration, the Kalman filter of theorem 5.5 should be compared with a simpler competitor, based on an extension of Luenberger's observers, which is now presented.

Generalizing Luenberger's definition, an observer for the hybrid system (5.2.1) is a dynamic system with state $\hat{x}_t \in \mathbf{R}^n$ described by

$$\dot{\hat{x}}_t = A^o(r_t)\, \hat{x}_t + K^o(r_t)\, y_t \tag{5.3.8}$$

for A^o and K^o, n x n and n x p, regime-dependent matrices such that $\hat{x}_{t_0} = x_{t_0}$ implies $\hat{x}_t = x_t$ for any $t > t_0$. Again the assumption of regime availability is explicitly used in (5.3.8) where the observer coefficients, A^o and K^o, are switched any time a jump occurs. The next theorem characterizes the choices for A^o and K^o that satisfy the condition $\hat{x}_t = x_t$, $t > t_0$, provided $\hat{x}_{t_0} = x_{t_0}$.

<u>Theorem 5.7</u>

Equation (5.3.8) defines a Luenberger's observer if

$$A^o(r_t) = A(r_t) - K(r_t)\, C(r_t)$$
$$K^o(r_t) = K(r_t)$$

for any arbitrary n x p matrices K_i, i = 1, N. The structure of the observer is then

$$\dot{\hat{x}}_t = A_i\, \hat{x}_t + K_i(y_t - C_i\, \hat{x}_t) \qquad \text{when } r_t = I$$

Proof:

For $\tilde{x}_t = x_t - \hat{x}_t$, the conditions of the theorem ensure that

$$\dot{\tilde{x}}_t = (A(r_t) - K(r_t)\, C(r_t))\, \tilde{x}_t$$

so that $\tilde{x}_{t_0} = 0$ ($\hat{x}_{t_0} = x_{t_0}$) indeed implies $\tilde{x}_t = 0$, $t > t_0$. □

The above result is not yet quite satisfactory because it uses an exact initialization while in practice x_{t_0} is rarely known (typically for the Kalman filter derivation x_{t_0} was assumed to be a gaussian random variable). An additional stability requirement is therefore imposed so that, starting from a non-zero error, $\tilde{x}_t$ decreases to zero as convergence of $\hat{x}_t$ to x_t occurs. Depending on the context, we may choose to impose either deterministic stability conditions, such as $\sigma(A_i - K_i\, C_i) < 0$, $i = 1, N$, or stochastic stability conditions involving A_i, K_i, C_i, $i = 1, N$ and Π (see chapters 2 and 3). It is important to note, however, that apart from this stabilization constraint, very few requirements are imposed on the K_is, which might lead us to consider the observer of theorem 5.7 as a valuable alternative to the optimal filter of theorem 5.5 when we do not want to solve N coupled Riccati equations.

Finally, another derivation of theorem 5.1 is proposed, exploiting the remaining degrees of freedom in the observer. With the above notations and using (5.3.8) as the state estimator, it results in an error variance X_i propagated as

$$\dot{X}_i = (A_i - K_iC_i)X_i + X_i(A_i - K_iC_i)' + F_i\, F_i' + K_iH_iH_i'K_i + \sum_{j=1}^{N} \pi_{ji}\, X_j \tag{5.3.9}$$

Introducing $K_i^* = X_iC_i'(H_i'H_i)^{-1}$, $i = 1, N$, we can write

$$(A_i - C_iK_i^*)X_i + X_i(A_i - C_iK_i^*) + K_i^*H_i'H_iK_i^* =$$
$$(A_i - C_iK_i)X_i + X_i(A_i - C_iK_i) + K_iH_i'H_iK_i - (K_i - K_i^*)H_i'H_i(K_i - K_i^*)$$

which shows that the right-hand side of (5.3.9), as a function of K_i, is minimal precisely for $K_i = K_i^*$. Hence the gains of the observer leading to the smallest error are the K_i^*s, for which (5.3.9) is precisely the set of coupled Riccati equations (5.3.5) of theorem 5.5.

The stability requirement for the observer is then satisfied under the conditions of theorem 5.6.

The Kalman filter of theorem 5.5 or any stable observers provide an estimate of the state variable from available measurements. A natural question is then how to use this reconstruction in a control loop. The answer is best formulated in terms of two central concepts of stochastic control theory, certainty equivalence and separation.

Consider a deterministic system (S_d) and a stochastic version of it (S_s) obtained for example by incorporating additive and multiplicative wide band noises in the plant state equation. For the deterministic system (S_d) some optimal control law might have been computed, maybe as a state feedback

$$u^* = U_d(x)$$

Going to the stochastic system (S_s), the state x is often no longer available but rather has to be estimated from the measurements, say as $\hat{x}_t = E\{x_t \mid Y_t\}$ where Y_t is some measurement σ-algebra. The optimal control law for (S_s) is in general very complex, e.g. it includes dual control effects. However, there are two special cases of interest:

- The Certainty Equivalence Principle is verified when the deterministic law can be used without alteration, simply replacing x by its estimate $\hat{x}$

$$u^* = u_{CEP} = U_d(\hat{x})$$

- The Separation Theorem is valid when the optimal law for the stochastic system uses the estimate $\hat{x}$ in a modified transformation

$$u^* = u_{ST} = U_s(\hat{x})$$

There is clearly a hierarchy between these two concepts, the Certainty Equivalence Principle being a very strong result since it states that the estimate $\hat{x}$ can be used exactly as if it were the true state x. Not surprisingly then the Certainty Equivalence Principle is valid only in rather restrictive situations. Alternatively these concepts can be used to define classes of control laws within which optimization for the stochastic model is performed. The separation theorem is attractive in this respect because it allows an independent design of the filter ($\hat{x}$) and of the regulator (U_s), which is a valuable simplification. In other words the designer often decides to enforce the separation theorem, that is, to look for the optimal solution within the class of control laws of the form $u = U(\hat{x})$ where $\hat{x}$ is the available state estimate.

The JLQG problem with the assumption of a measured regime is one of the few situations (as for the LQG sub-problem) where the certainty equivalence principle can be established.

Theorem 5.8:

The optimal solution of the JLQG problem with state and observation equations (5.2.1), (5.3.1) is the certainty equivalence solution

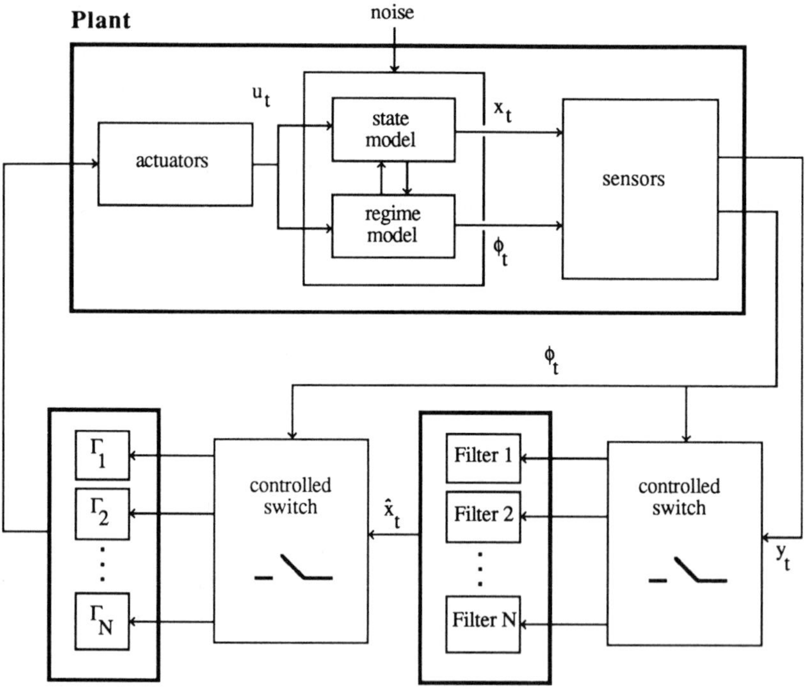

Figure 5.1- The Certainty Equivalence JLQG control law.

$$u_t^* = - Q_{2i}^{-1} B_i' \Lambda_i \hat{x}_t \qquad \text{when } r_t = i$$

where the Λ_is, i = 1 to N, are those of the piecewise deterministic JLQ regulator and where $\hat{x}_t$ is the optimal estimate of theorem 5.5.

Proof:

It is the same as the proof of the LQG result, see (Wonham 1971). □

The structure of the CEP solution and the role of the regime measurement channel is presented in figure 5.1 where the regime controls the switches for both the filter and the regulator.

For the filtering part of figure 5.1 there is one and only one Kalman filter running on-line at a given instant. This is to be contrasted with the bank of filters found in the multiple model literature (e.g. Lainotis 1976) where a number of filters run in parallel. The filter of theorem 5.5 is indeed a single filter with switching gains. It is of course the regime availability assumption which leads to simple structure and it will not be preserved in chapter 6 when the regime also has to be reconstructed from partial noisy observations.

5.4 Poisson Impulsive Disturbances

The Wiener process disturbances considered in §5.2, 5.3 represent an idealization of constantly acting unpredictable signals influencing the system behavior. In terms of the fault-tolerant control of a spacecraft discussed in chapter 1, the fluctuating perturbation torques due to gravity gradient sloshing residuals, etc. may be included in that category.

For this class of disturbances, and given the limited bandwidth of the system, Wiener or filtered Wiener processes are accepted as an adequate model. However, there is another kind of disturbance which manifests itself only at discrete random instants by an impulsive signal. Typically for the spacecraft example, this would correspond to the impingement of micrometeorites on the solar panels. Another mathematical model is then considered, the Poisson process.

The counterpart of (5.2.1) is

$$dx_t = A(r_t)\, x_t dt + B(r_t)u_t\, dt + D(x_t, r_t)\, d\pi_{1t} + E(u_t, r_t)d\pi_{2t} + F(r_t)d\pi_{3t} \quad (5.4.1)$$

where π_{it}, i = 1, 3, are independent Poisson processes on $\mathbf{R}^{n_i}$ with constant intensities λ_i, i = 1, 3, that are also independent of r_t and $x_{t_0} = x_0$ (see appendix 1 for a description of Poisson processes). Attention is restricted to the case where jump distribution $\psi_\ell(dz)$ of π_ℓ satisfies

$$\int_{\mathbf{R}^{n_\ell}} z\, \psi_\ell(dz) = 0\, ; \qquad \int_{\mathbf{R}^{n_\ell}} zz'\, \psi_\ell(dz) = Z_\ell(r_t)$$

that is, that the jump magnitude is a zero mean random variable with regime-dependent variance. Also, as in §5.2, D and E are assumed to be linear, in x and u respectively,

$$\begin{aligned} D(x_t, r_t) &= \sum_{k=1}^{n} x_{kt} D_k(r_t) \\ E(u_t, r_t) &= \sum_{k=1}^{m} u_{kt} E_k(r_t) \end{aligned} \qquad (5.4.2)$$

where x_{kt} (resp. u_{kt}) is the k-th component of x_t (resp. u_t). Next introduce endomorphisms **D** and **E** for P a symmetric n x n matrix as

$$\begin{aligned} \mathbf{d}_{k\ell}(r_t, P) &= \lambda_1 \mathrm{trace}(D_k(r_t)'P\, D_\ell(r_t)\, Z_1(r_t)) \qquad \mathbf{D} = (\mathbf{d}_{k\ell})_{k,\ell = 1,n} \\ \mathbf{e}_{k\ell}(r_t, P) &= \lambda_2 \mathrm{trace}(E_k(r_t)'P\, E_\ell(r_t)\, Z_2(r_t)) \qquad \mathbf{E} = (\mathbf{e}_{k\ell})_{k,\ell = 1,m} \end{aligned} \qquad (5.4.3)$$

With the above model, the optimal regulation problem of §5.2 is again considered. The first step is to compute the infinitesimal generator.

Theorem 5.9

For g(x, r) such that g and g_x are continuous in x for any r and such that $|\, g(x, r)\, | \leq \alpha(1 + |x|)$ for a constant α, the infinitesimal generator of the pair (x_t, r_t), for x_t solution of (5.4.1) and r_t a Markov chain on **S** with generator Π is given by

$$\mathscr{L}g(x, i) = -(\lambda_1 + \lambda_2 + \lambda_3)\, g(x, i) + \lambda_1 \int_{R^{n1}} g(x+D(x,i)z,i)\, \psi_1(dz)$$

$$+ \lambda_2 \int_{R^{n2}} g(x+E(u,i)z,i)\, \psi_2(dz) + \lambda_3 \int_{R^{n3}} g(x+F(x)z,i)\, \psi_3(dz) \qquad (5.4.4)$$

$$+ (A_i x + B_i u)' g_x(x, i) + \sum_{k=1}^{n} \pi_{ij} g(x,j)$$

Proof:

Repeat the proof of theorem 5.1. □

With this expression of the infinitesimal generator, dynamic programming gives the optimal regulator.

Theorem 5.10:

The control minimizing J for the system described by (5.4.1) through (5.4.3) is

$$u_t^* = -[E_i(\Lambda_i) + Q_{2i}]^{-1} B_i' \Lambda_i x \qquad \text{when } r_t = i$$

with the Λ_is i = 1, N, obtained from

$$-\dot{\Lambda}_i = A_i'\Lambda_i + \Lambda_i A_i - \Lambda_i B_i [Q_{2i} + E_i(\Lambda_i)]^{-1} B_i'\Lambda_i + Q_{1i} + \mathbf{D}_i(\Lambda_i) + \sum_{j=1}^{N} \pi_{ij} \Lambda_j$$

$$\Lambda_i(t_f) = 0 \qquad (5.4.5)$$

The minimal cost is

$$J^* = x_0' \Lambda_{i0}(t_0) x_0 + \mu_{i0}(t_0)$$

for $x_{t0} = x_0$ and $r_{t0} = i_0$, where the μ_is, i = 1, N, satisfy

$$-\dot{\mu}_i = \lambda_3 \text{ trace } (F_i'\Lambda_i F_i Z_{3i}) + \sum_{j=1}^{N} \pi_{ij} \mu_j$$

$$\mu_i(t_f) = 0 \qquad (5.4.6)$$

Proof:

As in the proof of theorem 5.2 we arrive at the Hamilton-Jacobi-Bellman equation in the cost-to-go V(x, i, t)

$$0 = \underset{u}{\text{Min}} \{V_t(x, r, t) + \mathscr{L}_u V(x, r, t) + x'Q_1x + u'Q_2u\}$$

Assuming V takes the form

$$V(x, i, t) = x_t'\Lambda_i x + \mu_i$$

the expression of $\mathscr{L}_u V$ is simplified from theorem 5.9

$$\mathscr{L}_u V(x, i, t) = (A_i x + B_i u)'\Lambda_i x + \sum_{j=1}^{N} \pi_{ij}(x_t'\Lambda_j x_t + \mu_j)$$
$$+ \frac{1}{2}[u'\mathbf{E}_i(\Lambda_i)u + x'\mathbf{D}_i(\Lambda_i)x + \lambda_3 \operatorname{trace}(F_i'\Lambda_i F_i Z_{3i})]$$

and from there onwards the proof proceeds as for theorem 5.2. □

The Poisson noises, π_i, i=1, 3, in (5.4.1) produce discontinuities of the state trajectory as explained above. However, they were assumed independent of the regime process and these discontinuities were intended to model the influence of impulsive exogeneous disturbances, such as the impact of micrometeorites on the solar panel of the spacecraft. It could happen, however, that the jumps of the regime engender discontinuities on their own, and indeed such a situation is studied in detail in chapter 7.

The situation where state discontinuities of random magnitude occur at the instant of regime jump can nevertheless already be discussed as a slight modification of theorem 5.10. Consider the model (5.4.1) with $\pi_1 = \pi_2 = 0$ and π_3 triggered by $c_t(i, j)$, the counting process recording the number of jumps from regime i to regime j up to time t. The intensity of $c_t(i, j)$ is obviously π_{ij}, the transition rate from i to j, i ≠ j. The random jump magnitude is now D(x, i, j)z with z distributed according to ψ with

$$\int_{R^{n3}} z\psi(dz) = 0 \ ; \qquad \int_{R^{n3}} zz'\psi(dz) = Z(r_t)$$

Assuming that D is linear in x

$$D(x, i, j) = \sum_{k=1}^{n} x_k D_k(i, j)$$

define accordingly the endomorphism **D**, for P a symmetric n x n matrix

$$\mathbf{d}_{k\ell}(i, j, P) = \mathrm{trace}(D_k(i, j)' P\, D_\ell(i, j) Z_i) \qquad \mathbf{D} = (\mathbf{d}_{k\ell})_{k,\ell=1,n}$$

for i≠j. The dependence of **D** on i and j is denoted as $\mathbf{D}_{ij}(P)$ and, by convention, $\mathbf{D}_{ii}(P) = 0$, i = 1, N.

The generator of (x_t, r_t) is adapted from theorem 5.9

$$\mathscr{L}g(x, i) = (A_i x + B_i u)' g_x(x, i) + \sum_{j=1}^{N} \pi_{ij}\, g(x, j) + \sum_{j=1}^{N} \pi_{ij} \int_{R^{n_3}} g(x+D(x,i,j)z, j)\, \psi(dz)$$

and a construction similar to that of theorem 5.2 and 5.10 leads to the optimal regulator

$$u_t^* = -\, Q_{2i}^{-1} B_i' \Lambda_i x_t \qquad \text{when } r_t = i$$

with

$$-\dot{\Lambda}_i = A_i'\Lambda_i + \Lambda_i A_i - \Lambda_i B_i Q_{2i} B_i'\Lambda_i + Q_{1i} + \sum_{j=1}^{N} \pi_{ij} (\Lambda_j + \Delta_{ij}(\Lambda_j))$$

$$\Lambda_i(t_f) = 0$$

5.5 Notes and References

Preserving for a while the assumption of an exact regime observation, this chapter has studied the influence of noises on JLQ regulators. The analysis covers the case of multiplicative as well as additive disturbances and also the two types of stochastic influences encountered in practice, continuous (Wiener) and discrete (Poisson).

A major finding is the certainty equivalence solution when only additive noises are considered. The resulting regulator has an appealing structure and could serve as a guideline for designing heuristic suboptimal solutions in more realistic situations.

The JLQG problem formulation owes much to (Wonham, 1971) with contributions to the LQG special case in (MacLane, 1970, 1971). The presence of multiplicative noises in the LQG setting was analyzed in (Mil'shtein, 1982, Phillis, 1985) and, when regime transition occurs, it has been shown in (Mariton, 1987a) that the Certainty Equivalence Principle is no longer valid. As an alternative, it was proposed that the matrices of a structurally constrained law be optimized by enforcing a separation theorem; that is, look for the optimal gains of the following regulator/filter cascade

$$d\hat{x}_t = A(r_t)\,\hat{x}_t dt + B(r_t)u_t\,dt + K(r_t)\,(dy_t - C(r_t)\,\hat{x}_t dt)$$

$$u_t = \Gamma(r_t)\,\hat{x}_t$$

The performance index being a combination of quadratic penalties in x_t, u_t and the estimation error $\tilde{x}_t$, necessary optimality conditions on the K_is, L_is, $i = 1, N$ are presented in (Mariton, 1987a) together with a qualitative analysis of the peculiar effects of multiplicative noises.

The occurrence of discontinuity of state trajectories was first introduced in (Sworder, 1972) using a deterministic counterpart of (5.4.1)

$$x_{t+} = \Gamma_{ij}\,x_t \quad \text{when } r_t = i,\ r_{t+} = j.$$

For the solar thermal receiver application of chapter 1, Sworder (1984) also studied the influence of constant discontinuities

$$x_{t+} = x_t + \alpha \quad \text{if the regime jumps at time t.}$$

Of course it is also possible to formulate constrained optimization problems for noisy hybrid dynamics. The optimal output feedback regulator with mode-dependent gains, for example, is given by the following theorem.

Theorem 5.11

A necessary condition for $u_t = \Gamma_i y_t$ when $r_t = i$ to be optimal for the JLQG problem is that the gains Γ_i, i = 1 to N, satisfy

$$\dot{X}_i = \tilde{A}_i X_i + X_i \tilde{A}_i' + \mathbf{D}_i(X_i) + F_i' F_i + \mathbf{E}_i(\Gamma_i C_i X_i C_i' \Gamma_i') + \sum_{j=1}^{N} \pi_{ji} X_j \, ; \quad X_i(t_0) = X_{i0}$$

$$-\dot{\Lambda}_i = \tilde{A}_i' \Lambda_i + \Lambda_i \tilde{A}_i + \mathbf{D}_i(\Lambda_i) + Q_{1i} + \mathbf{E}_i(\Gamma_i' C_i' \Lambda_i C_i \Gamma_i) + \sum_{j=1}^{N} \pi_{ij} \Lambda_j \, ; \quad \Lambda_i(t_f) = 0$$

$$\Gamma_i = -[Q_{2i} + \mathbf{E}_i(\Lambda_i)]^{-1} B_i' \Lambda_i X_i C_i' [C_i X_i C_i']^{-1}$$

where $\tilde{A}_i = A_i + B_i \Gamma_i C_i$, i = 1, N.

Proof:

Transform the problem to an equivalent deterministic one and apply the matrix maximum principle as for the constrained optimal regulators of chapter 3 (see also (Mariton, 1987b)). □

References

Athans, M., Ku, R., and Gerschwin, S.B. (1977). The uncertainty threshold principle: some fundamental limitations of optimal decision making under dynamic uncertainty. IEEE Trans. Aut. Control, AC-22: 491.

Bernstein, D.S., and Hyland, D.C. (1985). Optimal projection/maximum entropy stochastic modelling and reduced-order design synthesis, Proc. IFAC Workshop on Model Error Concepts and Compensation, Boston, pp.47-54.

Incertis, F.C. (1982). Optimal stochastic control of linear systems with state and control dependent noise: efficient computational algorithms, Nonlinear Stochastic Problems (R.S. Bucy and J.M.F. Moura, eds.), NATO/ASI Mathematical and Physical Sciences Series, n°104, p.243.

Lainotis, D.G. (1976). Partitioning: a unifying framework for adaptive systems, Part I-II, Proc. IEEE, 64: 1126, 1182.

MacLane, P.J. (1970). The optimal regulator problem for a stationary linear system with state dependent noise, ASME Trans. J. Basic Eng., 24: 363.

MacLane, P.J. (1971). Linear optimal estimation with state dependent disturbances and time correlated measurement noise, IEEE Trans. Aut. Control, AC-16: 198.

Mariton, M. (1987a). Joint estimation and control of jump linear systems with multiplicative noises, ASME Trans. J. Dyn. Systems Meas. Control, 109: 24.

Mariton, M. (1987b). Optimal output feedback for jump linear systems with state and control dependent noises, Control Theory Advanced Tech., 2: 633.

Mil'shtein, M. (1982). Design of stabilizing controller with incomplete state data for linear stochastic systems with multiplicative noise, Aut. Remote Control, 5: 653.

Phillis, Y.A. (1985). Controller design for systems with multiplicative noise, IEEE Trans. Aut. Control, AC-30: 1017.

Sworder, D.D. (1972). Control of jump parameter systems with discontinuous state trajectories, IEEE Trans. Aut. Control, AC-17: 740.

Sworder, D.D. (1984). Control of systems subject to small measurement disturbances, ASME Trans. J. Dyn. Systems Meas. Control, 106: 182.

Wonham, W.M. (1971). Random differential equations in control theory, Probabilistic Methods in Applied Mathematics (A.T. Bharucha-Reid, ed.) vol.2, Academic Press, New York, p.131.

6
Optimal Filtering

6.1 Changes in Signals and Systems

In chapter 5, we designed optimal filters and regulators for hybrid systems with a partially observed state. However a key assumption was that the discrete regime variable is available on-line, through a perfect observation channel. As was explained, the more realistic situation is when the regime information is polluted by some noise : in the solar thermal receiver application, the regime is a quantized version of the insolation, and photoelectric cells deliver a noisy measurement. The purpose of this chapter is to design filters that would extract in some optimal sense relevant information on the current regime of operation. Of course the Kalman filter in chapter 5 is recovered as a special case, when the noise on the regime measurement vanishes.

Some background on detection theory is provided in this introduction. The setting is as follows :

For some sample space Ω with a family of events Ξ, two hypotheses h and h_0 corresponding to probability measures $\mathcal{P}$ and $\mathcal{P}_0$ on (Ω, Ξ) are given, h associated with a normal regime and h_0 with a failed one, and we must decide which hypothesis is true on the basis of an observation process o_t. For the problem to be meaningful the statistics governing the observations must be different under h and h_0 and the purpose of detection theory is to define tests to reveal this difference. Denoting by O_t the (completed) σ-algebra generated by the observations $\{o_s, t_0 \leq s \leq t\}$, two approaches are now outlined. In the *bayesian approach*, it is assumed that an a priori measure on the set $\{h, h_0\}$ assigns probability p to h and p_0 to h_0. Two types of error are defined

$$p_e^{I} = \mathcal{P} \ \{ \text{deciding } h_0 \text{ is in force} \}$$

$$p_e^{II} = \mathcal{P}_0 \ \{ \text{deciding } h \text{ is in force} \}$$

called respectively the first and second kind errors. Denoting by $\mathscr{P}^t$ and $\mathscr{P}_0^t$ the restrictions of $\mathscr{P}$ and $\mathscr{P}_0$ to O_t and assuming that these are mutually absolutely continuous, the likelihood ratio of hypothesis h over h_0 is defined as the corresponding Radon-Nikodym derivative (see appendix 1)

$$\mathscr{L}_t = \frac{d\mathscr{P}^t}{d\mathscr{P}_0^t}$$

Restricting attention to threshold tests on the likelihood ratio

$$\text{decide} \quad \begin{array}{l} h \text{ when } \mathscr{L}_t(\omega) > \gamma \\ h_0 \text{ when } \mathscr{L}_t(\omega) < \gamma \end{array} \tag{6.1.1}$$

the total probability of error $pp_e^I + p_0 p_e^{II}$ is given by

$$p \int_{\mathscr{L}_t(\omega)<\gamma} \mathscr{P}\{d\omega\} + p_0 \int_{\mathscr{L}_t(\omega)>\gamma} \mathscr{P}_0\{d\omega\} \tag{6.1.2}$$

where the integrals are to be understood over the events such that $\mathscr{L}_t(\omega)$ is, respectively, smaller and greater than the threshold γ. But (6.1.2) can be written

$$p_0 + \int_{\mathscr{L}_t(\omega)<\gamma} [p\mathscr{L}_t(\omega) - p_0]\, \mathscr{P}_0\{d\omega\}$$

so that the optimal value of the threshold minimizing the total error is clearly $\gamma_0 = p_0 / p$. It can be shown (e.g. Ferguson, 1967) that this threshold test is actually optimal over a large class of decision strategies and bayesian performance criteria, see also (Davies, 1975). The simplicity of its structure makes it the test chosen in many applications, as shown in figure 6.1.

The crucial step is therefore to relate the likelihood ratio to the observations : in the off-line situation data are stored over $[t_0, t_f]$ and $\mathscr{L}_{t_f}$, is then computed from O_{t_f},

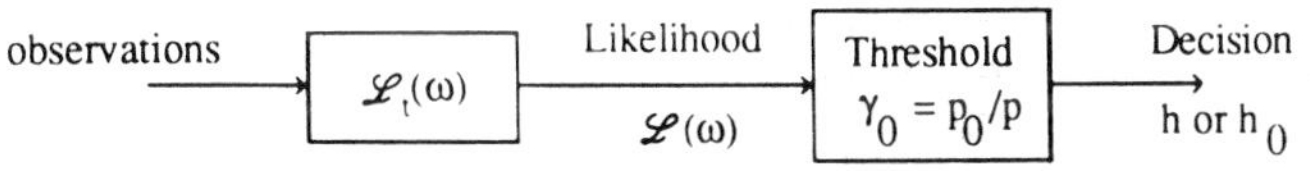

Figure 6.1 - Minimum probability of error bayesian detector.

while in the on-line situation it is desired to compute $\mathcal{L}_t$, possibly recursively based on O_t. The examples in chapter 1 indicate that the point of view to be privileged here is the on-line one but preventive maintenance in a fault tolerant control application could be cast as an off-line detection problem.

It is sometimes not possible to define an a priori measure on $\{h, h_0\}$ and the *Neyman-Pearson approach* is an alternative based on an interpretation of p_e^{II} as the probability of missed detection. To achieve accurate detection, the test must be sensitive to the high frequencies in the signal o_t that characterize a sudden change in the process, but as observations are in general noisy this unfortunately increases the possibility of false alarms. Clearly a compromise must be found and a rational way to do that is to look for the test minimizing p_e^{II} over all decision strategies achieving $p_e^{I} = \varepsilon$, for a given $\varepsilon > 0$. This is often called the most powerful test of level ε. The Neyman-Pearson lemma says : if there exists a number γ_1 such that $\mathcal{P}_0\ \{\omega \mid \mathcal{L}_t(\omega) \geq \gamma_1\} = \varepsilon$ then the test (6.1.1) with threshold γ_1 is the most powerful test of level ε.

In the bayesian setting optimal decision rules can be defined with respect to suitable risk functions, reflecting important features such as the cost of taking an additional measure, the cost of delaying a decision, or the cost of false alarms (Blackwell and Girschick, 1954).

As the markov model for the regime transitions constitutes an a priori description of the changes of hypothesis, our strategy is of the optimal bayesian type summarized by figure 6.1. To compute the required likelihood function in terms of the observations, we invoke the so-called separation of detection and filtering (Wong and Hajek, 1985): roughly speaking, the theorem states that the likelihood block $\mathcal{L}$ in figure 6.1 can be implemented as a cascade of a filter, estimating unknown variables, and

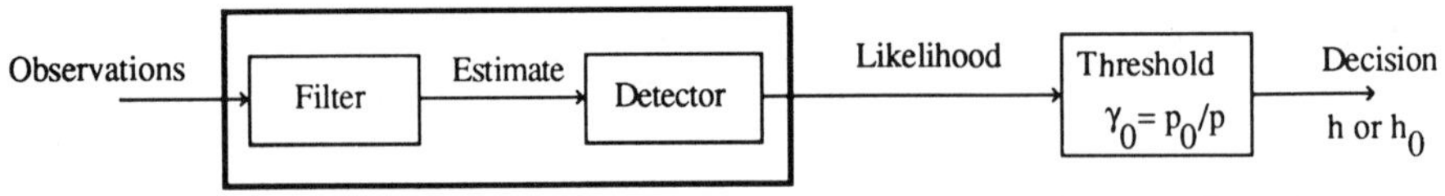

Figure 6.2 - Detection implementation.

detector, using the estimates *as if they were the exact variables* and computing the likelihood, as shown on figure 6.2.

In our regime detection problem, the test operations are particularly simple and attention is focused on the filtering problem, considering different observation models and information structures.

As in chapter 5, we shall use the a posteriori conditional expectation for our estimator. This is known to be optimal with respect to many estimation error cost function, in particular the celebrated mean square error criterion. We are therefore interested in computing quantities like $\hat{\phi}_t = E\{\phi_t \mid O_t\}$ based on the a priori jump model and some observations to be made precise later. Note that the ith component of $\hat{\phi}_t$ can be interpreted as the a posteriori probability of being in regime i at time t, $\hat{\phi}_{ti} = \mathcal{P}\{\phi_{ti} = 1 \mid O_t\}$.

6.2 Wiener Driven Observations

The simplest situation is considered first. It is motivated by the solar thermal receiver application of chapter 1 where photoelectric cells are present on the receiver panel to measure the incident flow of light. After some processing (averaging over the cells or selecting the signal corresponding to the highest flow) these cells deliver a signal which is a direct measure of the regime. This can be modeled by an observation equation

$$dy_t = h(r_t)\, dt + dv_t \tag{6.2.1}$$

where $h_1, h_2, \ldots, h_N$ represent the flow of light in the various regimes and v_t a measurement noise approximated by a brownian motion with intensity V. Using the regime indicator ϕ_t, an equivalent form of (6.2.1) is

$$dy_t = (h_1\ h_2 \ldots h_N)\phi_t dt + dv_t \tag{6.2.2}$$

with the jump dynamics of ϕ_t

$$d\phi_t = \Pi'\phi_t dt + dM_t \tag{6.2.3}$$

It is assumed that the Ξ-martingales v_t and M_t are independent. The observation σ-algebra is $Y_t = \sigma - \{y_s, t_0 \leq s \leq t\}$ and the optimal filter is given by the following theorem.

Theorem 6.1

The regime estimate $\hat{\phi}_t$ is generated by a set of stochastic differential equations

$$d\hat{\phi}_{ti} = (\Pi'\hat{\phi}_t)_i\, dt + \hat{\phi}_{ti}\, (h_i - \sum_{j=1}^{N} \hat{\phi}_{tj}\, h_j)\, V^{-1}\, (dy_t - \sum_{j=1}^{N} \hat{\phi}_{tj}\, h_j\, dt) \tag{6.2.4}$$

$i = 1, N$

from some initial value $\hat{\phi}_{t_0} = p_0$.

Proof :

From the fundamental result of optimal filtering, the estimation can be written

$$d\hat{\phi}_t = \Pi'\hat{\phi}_t dt + E\ \{(\phi_t - \hat{\phi}_t)\,(h_1\, h_2 \ldots h_N)\,(\phi_t - \hat{\phi}_t) \mid Y_t\}\ V^{-1}\,(dy_t - d\hat{y}_t)$$

With (6.2.2) and the martingale property of v_t it follows that

$$d\hat{y}_t = E\ \{(h_1\, h_2 \ldots h_N)\ \phi_t\, dt + dv_t \mid Y_t\}$$

$$= (h_1\, h_2 \ldots h_N)\ \hat{\phi}_t dt$$

$$= \sum_{j=1}^{N} \hat{\phi}_{tj}\, h_j\, dt$$

Next observe that

$$(\phi_t - \hat{\phi}_t)(h_1\, h_2 \dots h_N)(\phi_t - \hat{\phi}_t) = (\phi_t - \hat{\phi}_t)(\phi_t - \hat{\phi}_t)' \;(h_1\, h_2 \dots h_N)'$$

so that, using the usual notation $\tilde{\phi}_t = \phi_t - \hat{\phi}_t$ (see chapter 5 for example), the filter equation becomes

$$d\hat{\phi}_t = \Pi'\, \hat{\phi}_t\, dt + \widehat{\tilde{\phi}_t \tilde{\phi}_t'}\, (h_1\, h_2 \dots h_N)'\, V^{-1}\, (dy_t - \sum_{j=1}^{N} \hat{\phi}_{tj}\, h_j\, dt)$$

But ϕ_{ti} can take only values 0 or 1 and

$$\phi_t \phi_t' \;=\; \underset{i\,=\,1,N}{\text{diag}}\, (\phi_{ti}^2) = \underset{i\,=\,1,N}{\text{diag}}\, (\phi_{ti})$$

hence

$$\widehat{\phi_t \phi_t'} \;=\; \underset{i\,=\,1,N}{\text{diag}}\, (\hat{\phi}_{ti})$$

and the i-th component of the gain vector $\widehat{\tilde{\phi}_t \tilde{\phi}_t'}\, (h_1\, h_2 \dots h_N)'$ is obtained as

$$\hat{\phi}_{ti}\, h_i - \sum_{j=1}^{N} \hat{\phi}_{ti}\, \hat{\phi}_{tj}\, h_j = \hat{\phi}_{ti}\, (h_i - \sum_{j=1}^{N} \hat{\phi}_{tj}\, h_j)$$

which completes the proof of (6.2.4). At time t_0 the filter is initialized at p_0. □

It can be observed that the filter (6.2.4) has the familiar structure of the Kalman filter defined in chapter 5, combining an extrapolation on the prior model with measurement up-dates. It is simply a consequence of the fundamental optimal filter theorem which establishes the extrapolation/up-date structure in the most general setting. A major difference exists, however, regarding the up-date gain computation: in the classical (LQG) Kalman situation the gain can be precomputed off-line from the Riccati equation of the optimal error covariance. Here on the contrary the gain depends on the current estimate $\hat{\phi}_t$ and it must therefore be computed on-line. This leads to non-linearity in (6.2.4) with products $\hat{\phi}_{ti}\,\hat{\phi}_{tj}$, as opposed to the linearity of the Kalman filter equation, and implementation difficulties may result from this modulation between the estimates and the innovations $d\tilde{y}_t$ ($\tilde{y}_t = y_t - \hat{y}_t$). Some attention is given to this difficulty later on (§ 6.4).

In the absence of clues on the initial regime, the least presumptive choice is customarily made as $\hat{\phi}_{t_0 i} = \frac{1}{N}$, i = 1 to N.

To gain some insight into the behaviour of the filter (6.2.4) simulations are useful and the following example is considered

Example 1

For a two-regime system with $h_1 = 0.$ and $h_2 = 1.$, transition rates matrix $\Pi = \begin{pmatrix} -0.5 & 0.5 \\ 0.25 & -0.25 \end{pmatrix}$ and noise intensity V = 0.01, a sample path of $\hat{\phi}_{t1}$ is plotted on figure 6.3 when the true regime is one from $t_0 = 0$ to t = 300, two from t = 300 to t = 1000 and then back to one.

With the above values the signal to noise ratio ($\frac{(h_1 - h_2)^2}{V}$) is rather favourable . It can be observed that $\hat{\phi}_{t1}$ follows the path of ϕ_{t1} quite closely but also that there are some significant discrepancies between the two trajectories over limited periods of time. These are of two sorts : after a jump of ϕ_{t1} (at t = 300 and t = 1000 on figure 6.3) $\hat{\phi}_{t1}$ reacts only after a short lag which is called the detection delay, and there are also instances, see t = 130 or t = 850 , where $\hat{\phi}_{t1}$ moves, indicating a jump, while in truth ϕ_{t1} remained constant, this is called a false alarm.

The detection delay and the false alarm rate are the two key figures of merit in the detection literature and an algorithm is accepted if it produces a fast detection with few false alarms. To achieve short detection delays, the algorithm must respond quickly to changes in the last pieces of data, which implies a high "gain" at high frequencies and a sensitivity to noise generating false alarms. It is thus intuitive that small false alarm rates and short detection delays are contradictory objectives and some trade-off has to be decided. In our optimal filtering approach, the estimate $\hat{\phi}_t$ extracts information from the data in the best mean square sense and the trade-off is reported in the choice of a decision threshold. Looking at figure 6.3 a threshold $\hat{\phi}_{t1} \geq 0.8$ to decide a regime jump from 2 to 1

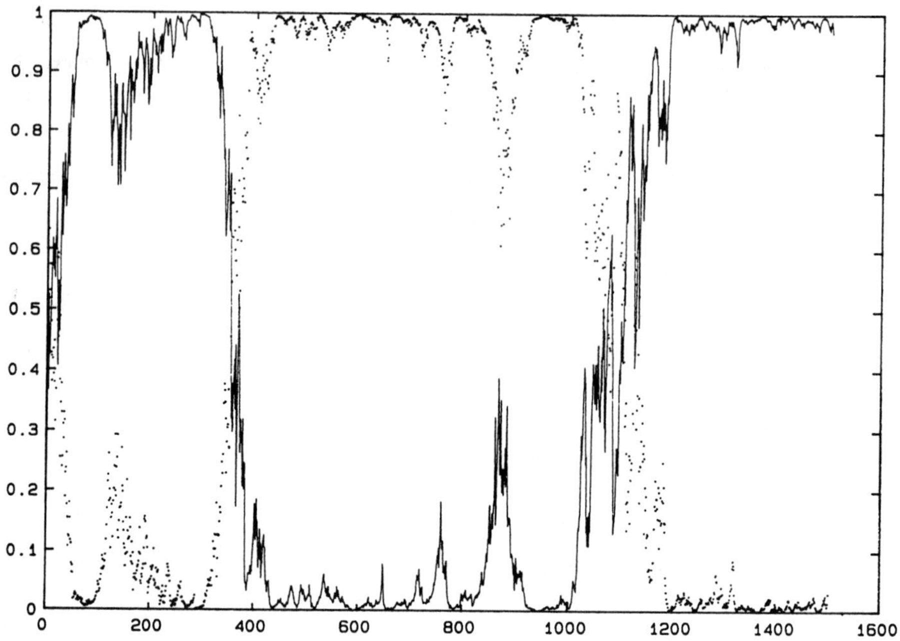

Figure 6.3 - Regime estimates.
(— $\hat{\phi}_{t1}$, ... $\hat{\phi}_{t2}$)

would give a detection at $t \approx 1150$ for a transition at $t = 1000$ and no false alarms. On the contrary a threshold 0.4 reduces the delay ($t_{detection} \approx 1050$) but creates a false alarm at $t \approx 900$.

The signal-to-noise ratio clearly influences the quality of the estimation and a less favourable situation is next considered

Example 2

The system now has three regimes with $h_1 = 0.$, $h_2 = 1.8$ and $h_3 = 2$. The transition rate matrix is chosen as $\Pi = \begin{pmatrix} -0.5 & 0.25 & 0.25 \\ 0.25 & -0.5 & 0.25 \\ 0.25 & 0.25 & -0.5 \end{pmatrix}$ with a noise intensity as before. The values h_i are such that the sensor clearly separates regime 1 from regime 2 or

3 but its perception of the difference between regime 2 and 3 is corrupted by the noise level. Figure 6.4 gives a sample path of $\hat{\phi}_{t1}$, $\hat{\phi}_{t2}$ and $\hat{\phi}_{t3}$ when the true regime is one up to t = 300 , two from t = 300 to 1000 and then three.

From the initial values $\hat{\phi}_{t_0 1} = \hat{\phi}_{t_0 2} = \hat{\phi}_{t_0 3} = 1/3$, the optimal filter quickly indicates regime 1 as the most plausible regime but after t = 300 when ϕ_{t2} jumps to 1, it is quite confused. It clearly perceives that 1 is no longer the true regime ($\hat{\phi}_{t1}$ drops near zero) but it cannot separate regime 2 and 3. After t = 1000 when ϕ_{t3} jumps to 1 the filter remains around $\hat{\phi}_{t2} \approx \hat{\phi}_{t3}$ for a while and then, accumulating data, slowly moves to favour regime 3. To improve the separation between regime 2 and 3 it is interesting to complement the regime measure (6.2.2) by some indirect link : for example when the state of the hybrid model is observed it contains an information on the current value of r_t, via the dependence of its dynamics. This solution is presented in theorem 6.2 below.

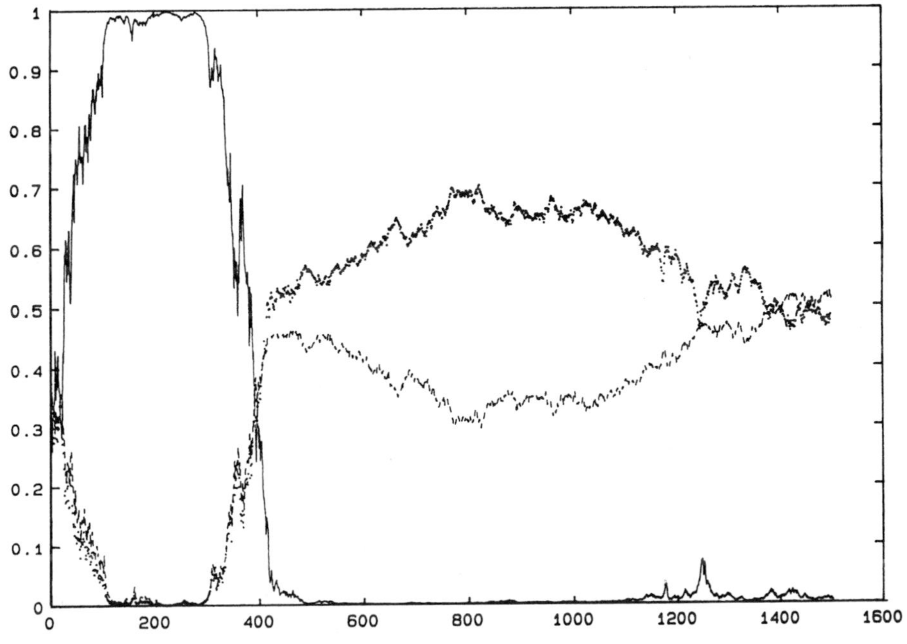

Figure 6.4 - Ambiguity of the estimator.
($—\hat{\phi}_{t1}$, --- $\hat{\phi}_{t2}$, ... $\hat{\phi}_{t3}$)

The above examples bring to light the role of signal-to-noise ratio, time and dynamics in the performance of the optimal filter estimating the regime changes. In fact this is related to the notion of consistency which is presented in § 6.4. Figure 6.4 shows that when the difference between two regimes ($h_2 - h_3$) is small compared to the noise intensity, then the optimal filter of theorem 6.1 has difficulty in separating these regimes. This signal-to-noise ratio also obviously influences the speed of convergence of the $\hat{\phi}_t$ trajectories.

The plant state contains information on the current regime via its dynamics which are now recalled.

$$dx_t = A(r_t)\, x_t\, dt + dw_t \tag{6.2.5}$$

with w_t a brownian motion of intensity W, independent of v_t. Note that it is assumed that x_t is measured and that the control influence ($B(r_t)\, u_t$) has been momentarily dropped. In what follows control terms could easily be included but as such they would not significantly enrich the situation described by (6.2.5). Interesting questions about the role of control concern the choice of u_t to accelerate convergence of the $\hat{\phi}_t$ estimates or simply to ensure consistency.

When the measurement channels (6.2.2) and (6.2.5) are available, the observation σ-algebra is $O_t = X_t \vee Y_t$ with $Y_t = \sigma\text{-}\{y_s, t_0 \le s \le t\}$ and $X_t = \sigma\text{-}\{x_s, t_0 \le s \le t\}$ and the corresponding optimal filter is given by

Theorem 6.2

The regime estimate $\hat{\phi}_t$ is generated by a set of stochastic differential equations

$$\begin{aligned} d\hat{\phi}_{ti} &= (\Pi'\hat{\phi}_t)_i\, dt + \hat{\phi}_{ti}\, (h_i - \sum_{j=1}^{N} \hat{\phi}_{tj}\, h_j)\, V^{-1}\, (dy_t - \sum_{j=1}^{N} \hat{\phi}_{tj}\, h_j\, dt) \\ &\quad + \hat{\phi}_{ti}\, x_t'\, (A_i - \sum_{j=1}^{N} \hat{\phi}_{tj}\, A_j)'\, W^{-1}\, (dx_t - \sum_{j=1}^{N} \hat{\phi}_{tj}\, A_j\, x_t\, dt) \end{aligned} \tag{6.2.6}$$

$i = 1, N$

from some initial value $\hat{\phi}_{t_0} = p_0$.

Proof :

The first two terms in (6.2.6) are obtained as in theorem 6.1 and, with the independence assumption, the last term comes from

$$E\,\{(\phi_t - \hat{\phi}_t)\,(A\,x_t - \hat{A x_t})'\,|\,O_t\}\;W^{-1}\;(dx_t - d\hat{x}_t)$$

But

$$d\hat{x}_t = (\sum_{j=1}^{N} \hat{\phi}_{tj}\,A_j)\;x_t\,dt$$

and

$$(\phi_t - \hat{\phi}_t)\,(A\,x_t - \hat{A x_t})' = (\phi_t - \hat{\phi}_t)\;x_t'\,(A - \hat{A})'$$

The expectation conditioned on X_t v Y_t gives for the i-th component of this N-dimensional vector

$$\hat{\phi}_{ti}\;x_t'\;(A_i - \sum_{j=1}^{N} \hat{\phi}_{tj}\,A_j)'$$

completing the proof. □

The structure of this filter resembles closely that of (6.2.4) and it is clear from the third term in (6.2.6) that the exact measurement of x_t is explicitly used.

Obviously if x_t is zero the filter (6.2.6) reduces to the previous one (6.2.4). Given that x_t is the deviation of the plant state x_{pt} from some desired level, it is likely that x_t is indeed close to zero during quiescent periods. However at the instants when a jump occurs an excitation of the linearized dynamics results from the regime dependence of the plant equilibria. The idea of using (6.2.5) excited at jump instants to improve regime estimation can be illustrated by going back to the previous example.

Example 2 revisited

In addition to y_t the filter now uses a scalar observation x_t generated by (6.2.5) with $A_1 = -0.5$, $A_2 = -5.$ and $A_3 = -50$. At jump times a discontinuity step is

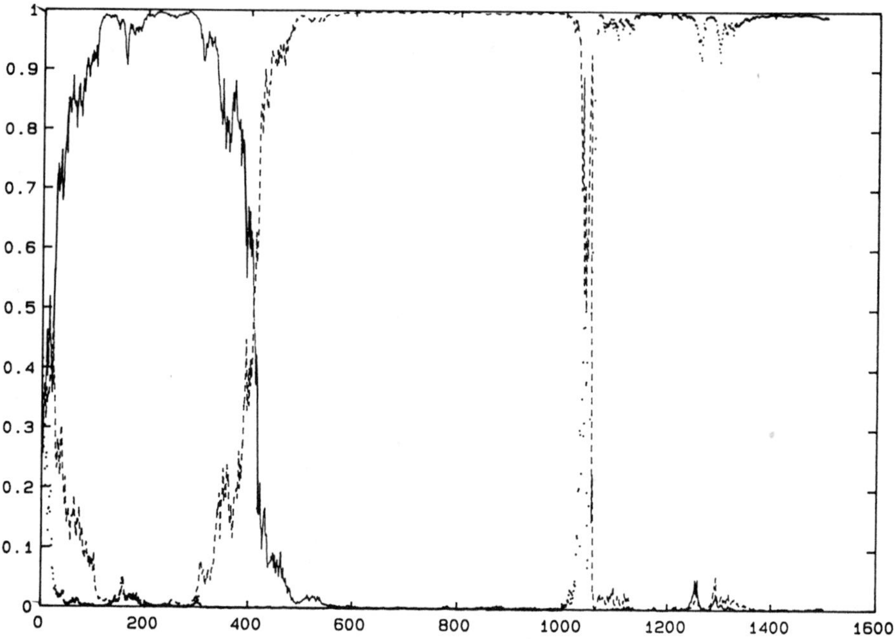

Figure 6.5 - Improved regime estimate.
($—\hat{\phi}_{t1}$, --- $\hat{\phi}_{t2}$, ... $\hat{\phi}_{t3}$)

artificially added to x_t (this procedure will be motivated in chapter 7). Figure 6.5 gives the $\hat{\phi}_t$ sample path with true jumps as in figure 6.4.

The improved separation of regimes 2 and 3 comes from the fact that $A_2 = -5.$ and $A_3 = -50.$ are more contrasted than $h_2 = 1.8$ and $h_3 = 2$. When excited at the instant of the jump from $r_t = 1$ to $r_t = 2$, the trajectory of x_t quickly influences $\hat{\phi}_t$ to indicate that a jump to regime 2 has occurred (we imposed $x = 0.05$ at $t = 300$ and $t = 600$). The corresponding sample path of x_t is plotted on figure 6.6.

The additional information on the regime contained in $A(r_t)\, x_t$ helps (6.2.6) distinguish regime 2 and 3 on figure 6.6 more accurately . Therefore this improvement is contingent upon the presence of an excitation in the system to disturb x_t from zero (recall that x_t is the deviation of the plant state x_{pt} from its nominal value x_n and the stabilization

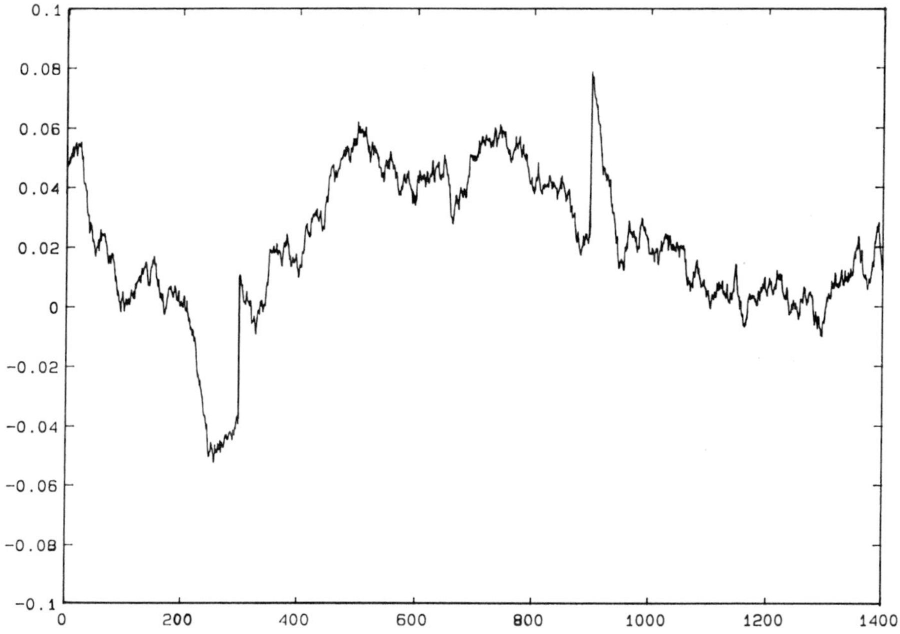

Figure 6.6 - State increment sample path (no control).

part of the control law is designed to regulated x_{pt} close to x_n). The state dynamics are controlled ($A(r_t)\, x_t + B(r_t)\, u_t$) and u_t may be used to excite the system thus giving rise to a dual control problem (see chapter 7). However this idea is not completely satisfactory since the temporary addition of a disturbing signal could only be decided from the $\hat{\phi}_t$ behaviour, and the initial transient of $\hat{\phi}_t$ after a jump would therefore not include this effect. Fortunately however the regime dependence of the plant equilibria ($x_n = x_n$ (i), i = 1 to N) introduces an excitation of the x_t dynamics.

We can also observe that non-linear dynamics would be easily accommodated in (6.2.6). For x_{pt} measured, only the last term of (6.2.6) would be modified to become

$$\hat{\phi}_{ti}\, [f(x_{pt}, u_{pt}, i) - \sum_{j=1}^{N} \hat{\phi}_{tj}\, f(x_{pt}, u_{pt}, j)]\; W^{-1}[dx_{pt} - \sum_{j=1}^{N} \hat{\phi}_{tj}\, f\,(x_{pt}, u_{pt}, j)\, dt]$$

where we have used the notation f(-, -, i) for f (-, -, ϕ_t) with $\phi_{ti} = 1$.

Of course the filter of theorem 6.2 is based on an unrealistic assumption because in general, only some components of x_t are observed. In the rest of this section the observation (6.2.2) is dropped and we consider, for w_t a brownian motion with intensity W,

$$dx_t = A(r_t)\, x_t dt \; + dw_t$$
$$d\phi_t = \Pi' \phi_t dt \; + dM_t \tag{6.2.7}$$

with measurements

$$dz_t = C(r_t)\, x_t dt + dv_t \tag{6.2.8}$$

where v_t is again an independent brownian motion with intensity V, z_t is an observation vector ($z_t \in \mathbf{R}^p$) and the C_i's, i = 1 to N, p x n matrices. This situation corresponds for example to the target tracking application of chapter 1 when the manoeuvres of the target through sudden changes in accelerations ($A = A(r_t)$) are to be estimated and detected using partial measurement ($p \leq n$), of the state.

The observation σ-algebra is now $Z_t = \sigma - \{z_s, t_0 \leq s \leq t\}$ and the optimal filter estimates simultaneously the regime ($\hat{\phi}_t = E\{\phi_t \mid Z_t\}$) and the state ($\hat{x}_t = E\{x_t \mid Z_t\}$).

<u>Theorem 6.3</u>

The regime estimate $\hat{\phi}_t$ is generated by a set of stochastic differential equations

$$d\hat{\phi}_{ti} = (\Pi' \hat{\phi}_t)_i dt + \hat{\phi}_{ti} \, (C_i \hat{x}_{ti} - \sum_{j=1}^{N} \hat{\phi}_{tj} C_j \hat{x}_{tj})' \, V^{-1} \, (dz_t - \sum_{j=1}^{N} \hat{\phi}_{tj} C_j \hat{x}_{tj} dt) \tag{6.2.9}$$

with $\hat{\phi}_{t_0} = p_0$ some initial regime distribution. The required $\hat{x}_{ti}$, i = 1 to N, are related to the state estimate

$$\hat{x}_t = \sum_{i=1}^{N} \hat{\phi}_{ti} \hat{x}_{ti} \tag{6.2.10}$$

and they are themselves propagated as

$$d\hat{x}_{ti} = A_i\hat{x}_{ti}dt + \frac{1}{\hat{\phi}_{ti}}\sum_{j=1}^{N} \pi_{ji}\ (\hat{x}_{tj} - \hat{x}_{ti})\ \hat{\phi}_{tj}\ dt + X_{ti}\ C_i^{'}\ V^{-1}(dz_t - C_i\hat{x}_{ti}\ dt) \tag{6.2.11}$$

with $X_{ti} = E\{(x_t - \hat{x}_{ti})\ (x_t - \hat{x}_{ti})' \mid Z_t, r_t = i\ \}$ and some initial value $\hat{x}_{t0i} = \hat{x}_{0i}$.

<u>Proof</u> :

The first term in (6.2.9) is the usual propagation term based on the a priori model for ϕ_t transitions. The gain for the innovation $dz_t - d\hat{z}_t = dz_t - \sum_{j=1}^{N} \hat{\phi}_{tj}\ C_j\ \hat{x}_{tj}\ dt$ is

$$E\ \{(\phi_t - \hat{\phi}_t)\ (Cx_t - \hat{Cx}_t)' \mid Z_t\}$$

the state x_t is not observed but we compute

$$\hat{Cx}_t = E\ \{C(r_t)\ x_t \mid Z_t\} = E\ \{\sum_{j=1}^{N} \phi_{tj}\ C_j\ x_t \mid Z_t\}$$

Defining

$$\hat{x}_{ti} = E\ \{x_t \mid Z_t, r_t\ = i\ \}$$

we can write

$$\hat{Cx}_t = \sum_{j=1}^{N} \hat{\phi}_{tj}\ C_j\ \hat{x}_{tj}$$

and the gain is computed as in theorem 6.2

$$E\ \{(\phi_t - \hat{\phi}_t)\ (Cx_t - \hat{Cx}_t)' \mid Z_t\} = \hat{\phi}_t\ \ (C_i\ \hat{x}_{ti} - \sum_{j=1}^{N} \hat{\phi}_{tj}\ C_j\hat{x}_{tj})$$

which completes the proof of (6.2.9). Equation (6.2.10) comes from

$$\hat{x}_t = E\{x_t \mid Z_t\} = E\{\sum_{i=1}^{N} x_t \phi_{ti} \mid Z_t\}$$

but $\phi_{ti}^2 = \phi_{ti}$ and

$$E\{\sum_{i=1}^{N} x_t \phi_{ti} \mid Z_t\} = \sum_{i=1}^{N} E\{\phi_{ti} x_t \phi_{ti} \mid Z_t\} = \sum_{i=1}^{N} \hat{\phi}_{ti} \hat{x}_{ti}$$

Regarding the $\hat{x}_{ti}$ filter, it is convenient to introduce the auxiliary variable $\xi_{ti} = x_t \phi_{ti}$. It obeys

$$d\xi_{ti} = \phi_{ti} (A(r_t) x_t \, dt + dw_t) + (\Pi' \phi_t)_i \, x_t dt + x_t \, dM_{ti} \tag{6.2.12}$$

By the independence assumption the co-variation between w_t and M_t vanishes. Now $\hat{\xi}_{ti} = E\{\xi_{ti} \mid Z_t\}$ is obtained as

$$d\hat{\xi}_{ti} = \hat{\phi}_{ti} A_i \hat{x}_{ti} \, dt + \sum_{j=1}^{N} \pi_{ji} \hat{\phi}_{tj} \hat{x}_{tj} \, dt + E\{(\xi_{ti} - \hat{\xi}_{ti})(\xi_{ti} - \hat{\xi}_{ti})' \mid Z_t\} C_i' V^{-1}(dz_t - C_i \hat{x}_{ti} \, dt)$$

From $\hat{\xi}_{ti}$, the desired $\hat{x}_{ti}$, i = 1 to N, are deduced using $\hat{x}_{ti} = \hat{\xi}_{ti} / \hat{\phi}_{ti}$ and Ito's rule for chain derivation with independent w_t and M_t which gives

$$d\hat{x}_{ti} = -\frac{1}{\hat{\phi}_{ti}^2} \hat{\xi}_{ti} \, d\hat{\phi}_{ti} + \frac{1}{\hat{\phi}_{ti}} d\hat{\xi}_{ti}$$

leading to

$$d\hat{x}_{ti} = -\frac{1}{\hat{\phi}_{ti}} \left(\sum_{j=1}^{N} \pi_{ji} \hat{\phi}_{tj}\right) \hat{x}_{ti} \, dt + \frac{1}{\hat{\phi}_{ti}} \left(\hat{\phi}_{ti} A_i \hat{x}_{ti} \, dt + \sum_{j=1}^{N} \pi_{ji} \hat{\phi}_{tj} \hat{x}_{tj} \, dt\right)$$

$$+ \frac{1}{\hat{\phi}_{ti}} E\{(\xi_{ti} - \hat{\xi}_{ti})(\xi_{ti} - \hat{\xi}_{ti})' \mid Z_t\} C_i' V^{-1} (dz_t - C_i \hat{x}_{ti} \, dt)$$

Grouping the two summations on j, this gives (6.2.11) of the theorem with a gain factor

$$\frac{1}{\hat{\phi}_{ti}} E\{(\xi_{ti} - \hat{\xi}_{ti})(\xi_{ti} - \hat{\xi}_{ti})' \mid Z_t\} = E\{(x_t - \hat{x}_{ti})(x_t - \hat{x}_{ti})' \mid Z_t, r_t = i\}$$

which is precisely the definition of X_{ti}, thus completing the proof. □

Contrary to the situation of theorems 6.2 and 6.1 the optimal filter is not completely defined by (6.2.9) to (6.2.11) and an equation for the conditional variance X_{ti} is needed. As in the proof define

$$\underline{\xi}_{ti} = \phi_{ti}\, x_t\, x_t' - \hat{\phi}_{ti}\, \hat{x}_{ti}\, \hat{x}_{ti}'$$

A direct computation shows that $\hat{\underline{\xi}}_{ti} = \hat{\phi}_{ti} X_{ti}$ and the search for X_{ti} can be replaced by that for $\hat{\underline{\xi}}_{ti}$. By repeated use of Ito's differentiation rule, we find that the terms of $d\underline{\xi}_{ti}$ contributing to $d\hat{\underline{\xi}}_{ti}$ are

$$(\Pi'\phi_t)_i\, x_t x_t'\, dt + \phi_{ti}\, d(x_t x_t') + \phi_{ti}(A_i x_t x_t' + x_t x_t' A_i')dt + \phi_{ti}\, dw_t\, dw_t'$$

$$- \sum_{j=1}^{N} \pi_{ji}\, \hat{\phi}_{tj}\, dt\, \hat{x}_{ti}\, \hat{x}_{ti}' - \hat{\phi}_{ti}(A_i \hat{x}_{ti} \hat{x}_{ti}' + \hat{x}_{ti} \hat{x}_{ti}' A_i')dt$$

$$- \hat{\phi}_{ti}\Big[\frac{1}{\hat{\phi}_{ti}}\Big(\sum_{j=1}^{N} \pi_{ji}\, \hat{\phi}_{tj}(\hat{x}_{tj} - \hat{x}_{ti})\Big)\hat{x}_{ti}' + \hat{x}_{ti}\frac{1}{\hat{\phi}_{ti}}\Big(\sum_{j=1}^{N} \pi_{ji}\, \hat{\phi}_{tj}(\hat{x}_{tj} - \hat{x}_{ti})\Big)'\Big]dt \tag{6.2.13}$$

$$- \hat{\phi}_{ti}\, X_{ti}\, C_i'\, V^{-1} C_i\, X_{ti}\, dt$$

The estimate of the auxiliary variable $\underline{\xi}_{ti}$ is then obtained from (6.2.13) by the filter representation theorem. We find

$$d\hat{\underline{\xi}}_{ti} = \hat{\phi}_{ti}(W - X_{ti} C_i' V^{-1} C_i X_{ti} + A_i X_{ti} + X_{ti} A_i')dt$$

$$+ \Big[\sum_{j=1}^{N} \pi_{ji}\, \hat{\phi}_{tj}(\widehat{x_t x_t'\phi_{tj}} - \hat{x}_{ti} \hat{x}_{ti}' - (\hat{x}_{tj} - \hat{x}_{ti})\, \hat{x}_{ti}'\Big]dt$$

(6.2.14)

$$- \hat{x}_{ti} [\sum_{j=1}^{N} \pi_{ji} \hat{\phi}_{tj} (\hat{x}_{tj} - \hat{x}_{ti})']dt$$

$$+E \{(\xi_{ti} - \hat{\xi}_{ti}) (\xi_{ti} - \hat{\xi}_{ti}) \mid Z_t\} C_i' V^{-1}(dz_t - C_i \hat{x}_{ti} dt)$$

Going back to X_{ti}, (6.2.14) is transformed into

$$dX_{ti} = (A_i X_{ti} + X_{ti} A_i' + W - X_{ti} C_i' V^{-1} C_i X_{ti})dt$$

$$+ \frac{1}{\hat{\phi}_{ti}} [\sum_{j=1}^{N} \pi_{ji} \hat{\phi}_{tj} (X_{tj} - X_{ti} + (\hat{x}_{tj} - \hat{x}_{ti}) (\hat{x}_{tj} - \hat{x}_{ti})')] dt \tag{6.2.15}$$

$$+ E \{\tilde{x}_{ti} \tilde{x}_{ti}' \tilde{x}_{ti} \mid Z_t, r_t = i\} C_i' V^{-1}(dz_t - C_i \hat{x}_{ti} dt)$$

The first term of this expression is the familiar Riccati equation of linear gaussian filtering problems. However, as for $\hat{x}_{ti}$, jumps induce a coupling which is given by the second term, and, more important, the second moment equation is driven by the innovations through the last term. The gain is given by the third moment of the error and to compute (6.2.15) its value is needed.

Proceeding as for the second moment we could write an equation for the third moment but this would include an innovation driving term with the fourth moment modulating the gain. The difficulty is now clear : to compute the estimate (= first moment) we need the second moment but to compute the second moment we need the third moment and so on. This situation is often summarized by saying that the moment equations are not closed or that the optimal filter is infinite dimensional. In the usual Kalman filter, it is the gaussianness of the estimation error $x_t - \hat{x}_t$ which closes the moment equations by ensuring that the third moment is zero.

6.3 Point Process Observations

There are practical situations where information on the current regime is more readily described in terms of a point process. Consider for example a manufacturing system operating under two regimes, say a nominal regime $r_t = 1$ and a failure regime $r_t = 2$. A specificity of manufacturing systems, as opposed to batch processes familiar in the chemical industry, is that their outputs are intrinsically discrete, like parts in a car factory or chips in an electronics assembly line. Very often some tests, maybe under a random sampling, are performed on the products delivered by the manufacturing system and, depending on the client requirements, are classified as acceptable or rejected and dumped. The observations can thus be aggregated into counting processes, one per category of the classification test. If products are marked either as ok or as rejected, this leads to two counting processes, one counting the number of products that have passed the test up to the current time and the other recording the number of rejected products. These counting processes are dependent upon the regime in the sense that the base rate for accepted products is higher in the normal regime and lower when machine operation is degraded.

Another example is provided by the detection of incidents on freeways. Using imaging sensors (typically a video camera associated with some pattern recognition algorithm) a lane by lane classification of vehicles is obtained according to their speed. The flow of vehicles being modeled as a Poisson process, we again find a situation where information on the presence of an incident on a given section is contained in the rate of the point process (obviously a larger rate is attributed to the normal regime, say $r_t = 1$, than to the incident $r_t = 2$) and in its marks (high speeds for $r_t = 1$ and low speeds for $r_t = 2$).

In both examples there is a possibility of misclassification, a failure of the quality test itself for the manufacturing system or a wrong speed when trucks and cars are mixed on a freeway. Our model must therefore recognize that the counter recording a given mark is sometimes incremented by falsely appropriating an event associated with another mark.

The above situations can be formulated mathematically as follows : for a hybrid system with N regimes and transitions described by a Markov chain as previously

$$d\phi_t = \Pi'\phi_t + dM_t$$

we introduce N counting processes $z_t(1)$ to $z_t(N)$ with rates

$$\lambda^i = \lambda_1 d_{i1} \phi_{t1} + \lambda_2 d_{i2} \phi_{t2} + ... + \lambda_N d_{iN} \phi_{tN} \qquad i = 1, N \qquad (6.3.1)$$

where $D = (d_{ij})_{i,j=1,N}$ is a matrix of discernability probabilities. The interpretation of (6.3.1) is that when the regime is j ($\phi_{tj} = 1$) the counter number i is incremented with a poisson rate $\lambda_j d_{ij}$ with λ_j the base rate corresponding to the current regime operation. The quality of the classification test should be reflected in the diagonal dominant character of the discernability matrix D. Writing $\lambda^i(\phi_t)$ for (6.3.1) the observation equations are summarized by

$$dz_t(i) = \lambda^i(\phi_t)\, dt + d\mu_t^i \qquad i = 1 \text{ to } N$$

where the μ_t^is are Ξ-martingales which are assumed independent of M_t the regime martingale. Denoting by Z_t the σ-algebra generated by the counting observations, $Z_t = \sigma$ - $\{z_s(i), t_0 \le s \le t, i = 1,N\}$, the optimal regime estimate $\hat{\phi}_t = E\{\phi_t \mid Z_t\}$ is given by the next theorem, assuming that the rates satisfy the boundedness and non degeneracy conditions recalled in appendix 1.

<u>Theorem 6.4</u>

The regime estimate $\hat{\phi}_t$ is generated by a set of stochastic differential equations

$$d\hat{\phi}_{ti} = (\Pi'\hat{\phi}_t)_i\, dt$$

$$+ \sum_{k=1}^{N} \frac{1}{\sum_{j=1}^{N} \lambda_j d_{kj} \hat{\phi}_{tj}^-} [\lambda_i d_{ki} - \sum_{j=1}^{N} \lambda_j d_{kj} \hat{\phi}_{tj}^-]\, \hat{\phi}_{ti}^- [dz_t(k) - \sum_{j=1}^{N} \lambda_j d_{kj} \hat{\phi}_{tj}\, dt]$$

$i = 1,N$ (6.3.2)

with $\hat{\phi}_{t_0} = p_0$ some initial regime distribution.

<u>Proof</u> :

Thanks to the independence hypothesis, it is sufficient to consider one term in the summation on k to prove (6.3.2), the extrapolation term $\Pi'\hat{\phi}_t$ resulting as before from the a priori Markov model of ϕ_t.

We extract the discontinuous part of the optimal filter representation theorem (see appendix 1) to find that up-dates on the k-th counter $z_t(k)$ are performed according to

$$\frac{1}{E\{\lambda_t^k \mid Z_t^-\}} \; E\{(\phi_t - \hat{\phi}_t)(\lambda_t^k - \hat{\lambda}_t^k) \mid Z_t^-\} \, [dz_t(k) - \hat{\lambda}_t^k dt]$$

where λ^k is given by (6.3.1) and E { - | Z_t^-} gives the predictable version of the estimate, noted more conveniently $\hat{\hat{\lambda}}$. From (6.3.1) it is seen that

$$\hat{\lambda}_t^k = \sum_{j=1}^{N} \lambda_j \, d_{kt} \, \hat{\phi}_{tj}$$

and

$$\hat{\hat{\lambda}}_t^k = \sum_{j=1}^{N} \lambda_j \, d_{kt} \, \hat{\phi}_{t\bar{j}}$$

The i-th component of the vector $(\phi_t - \hat{\phi}_t)(\lambda_t^k - \hat{\lambda}_t^k)$ is treated as in the proof of theorem 6.1 to produce

$$\hat{\phi}_{t\bar{i}} \; [\lambda_i \, d_{ki} - \sum_{j=1}^{N} \lambda_j \, d_{kj} \, \hat{\phi}_{t\bar{j}}]$$

Putting these three terms together then gives (6.3.2) of the theorem. □

If there is a correlation between the regime martingale M_t and the observation noise martingales μ_t^i, i = 1 to N, the gain in (6.3.2) has to be modified with the covariation term $< M, \mu^i >$ as indicated in appendix 1.

It is of course possible to use simultaneously continuous and discontinuous observations and, provided the observation martingales are independent, the optimal filter is obtained by adding to (6.3.2) the up-date term of theorem 6.1 or 6.2. In the manufacturing system example, this is indeed interesting because machines are usually equipped with sensors monitoring, say, the vibration level or the power consumption, and a change in vibration level or an increase of power consumption might signal a regime transition. It is then interesting to fuse in a single filter the information provided by these continuous observations with the information contained in the point process delivery of parts with marked qualities.

The up-dates in (6.3.2) occur at random instants and the base rates obviously influence the dynamics of $\hat{\phi}_t$. To illustrate this influence qualitatively the following example is studied.

Example :

Consider a system with three regimes and transition rates matrix

$$\Pi = \begin{pmatrix} -0.5 & 0.25 & 0.25 \\ 0.25 & -0.5 & 0.25 \\ 0.25 & 0.25 & -0.5 \end{pmatrix}.$$

Two transitions are considered, r_t jumps from 1 to 2 at t = 300 and to 3 at t = 600. The discernability matrix is $D = \begin{bmatrix} 0.8 & 0.1 & 0.1 \\ 0.1 & 0.8 & 0.1 \\ 0.1 & 0.1 & 0.8 \end{bmatrix}$ and the base rates are all equal to λ =

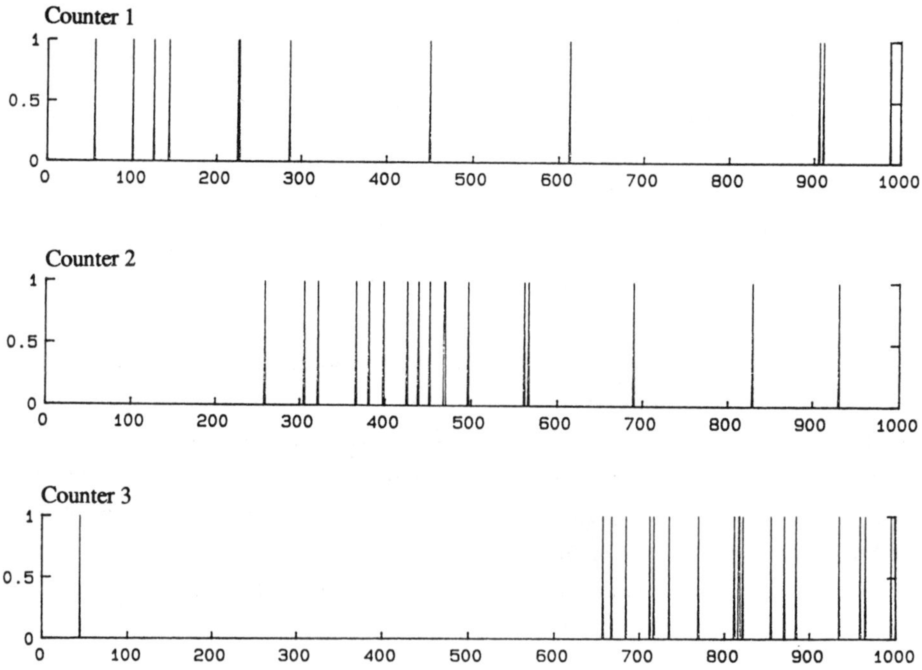

Figure 6.7 - Counting process observations.

400. Figure 6.7 presents a sample path of the three corresponding counting processes, with dz_t (1) on the top and dz_t (3) at the bottom

The corresponding $\hat{\phi}_{ti}$, i = 1, 3, are on figure 6.8

We observe a good separation of the regimes with discrete up-dates and exponential relaxation when the filter is not excited by new data. The influence of the base rate is shown by increasing λ to 2000 so that there are, on the average, five times more counts per unit of time. The sensors increments are plotted on figure 6.9 with again $dz_t(1)$ above $dz_t(2)$ and $dz_t(3)$ and the corresponding estimates are given on figure 6.10

With the increased rate convergence occurs very quickly. However in the limit of a very high rate it is well known that we simply recover brownian observations as in § 6.2. This is illustrated by the similarity with the estimates of §6.2 observed on figure 6.11 below where λ = 3 000 and $r_t = 2$ up to t = 300, $r_t = 1$ between t = 300 and t = 1000 and then back to $r_t = 2$. We used matrices

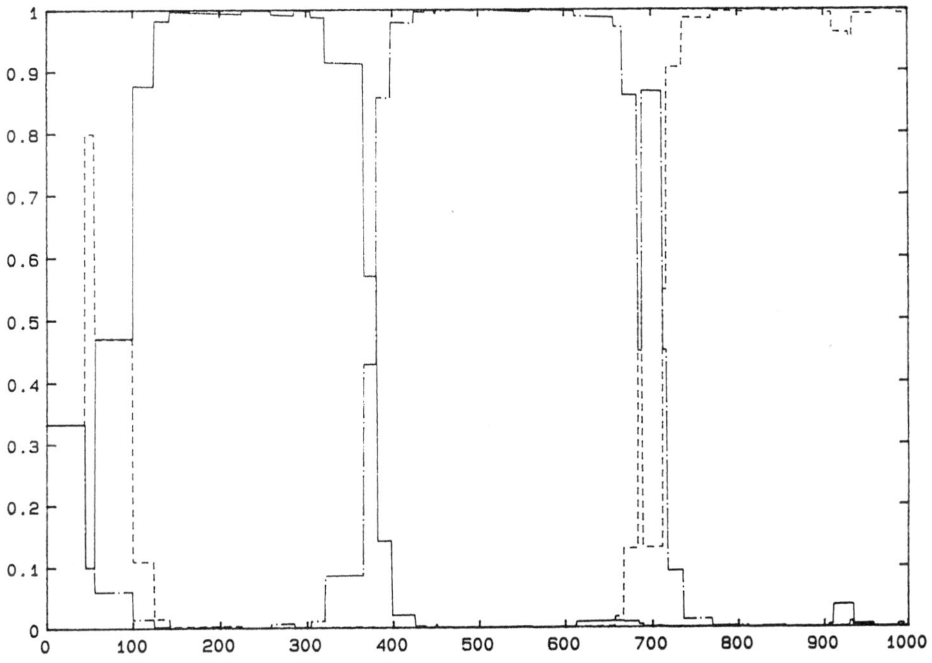

Figure 6.8 - Regime estimate trajectories.

($—\hat{\phi}_{t1}$, $-\cdot-\hat{\phi}_{t2}$, --- $\hat{\phi}_{t3}$)

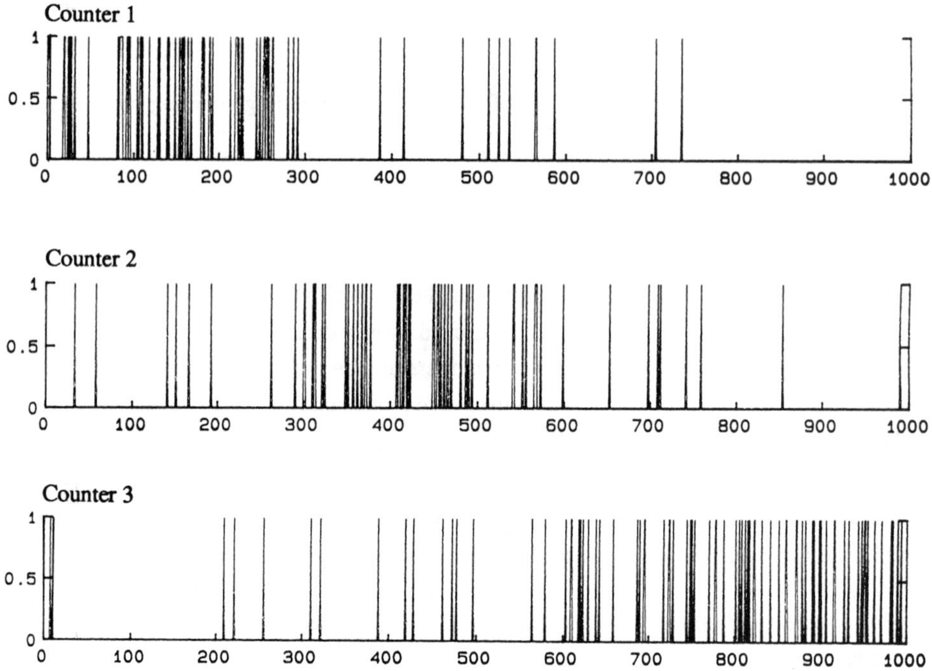

Figure 6.9 - Counting observations with increased rate.

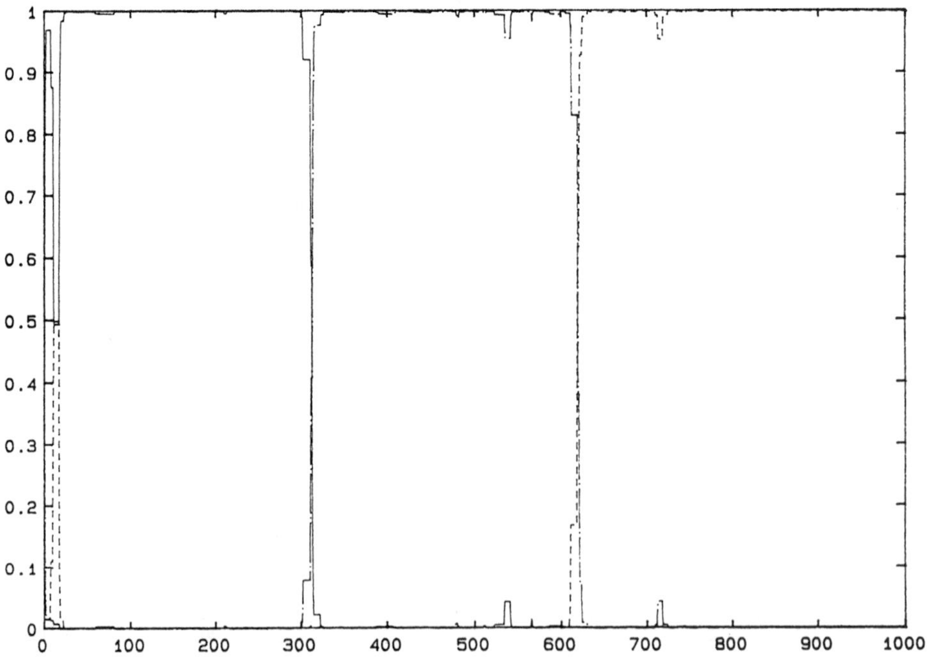

Figure 6.10 - Regime estimate trajectories. (— $\hat{\phi}_{t1}$, – - – $\hat{\phi}_{t2}$, --- $\hat{\phi}_{t3}$)

$$\Pi = \begin{pmatrix} -1 & 0.8 & 0.2 \\ 0.15 & -0.3 & 0.15 \\ 0.8 & 0.2 & -1 \end{pmatrix}$$

and

$$D = \begin{pmatrix} 0.5 & 0.1 & 0.4 \\ 0.1 & 0.8 & 0.1 \\ 0.4 & 0.1 & 0.5 \end{pmatrix}$$

The value of D indicates that regimes 1 and 3 are hard to distinguish. When r_t jumps to 1 the filter quickly detects that we left r = 2 but it is not clear whether the destination was r = 1 or r = 3. This is because in D we have indicated that when r = 1 (or 3) there is one chance out of two (0.1 + 0.4) that a misclassification occurs. Also a false alarm occurs around t = 900 because the Π matrix corresponds to a short average lifetime in regime 1 and the filter expects a transition out of r = 1 and tends to see one in noise, generating false alarms.

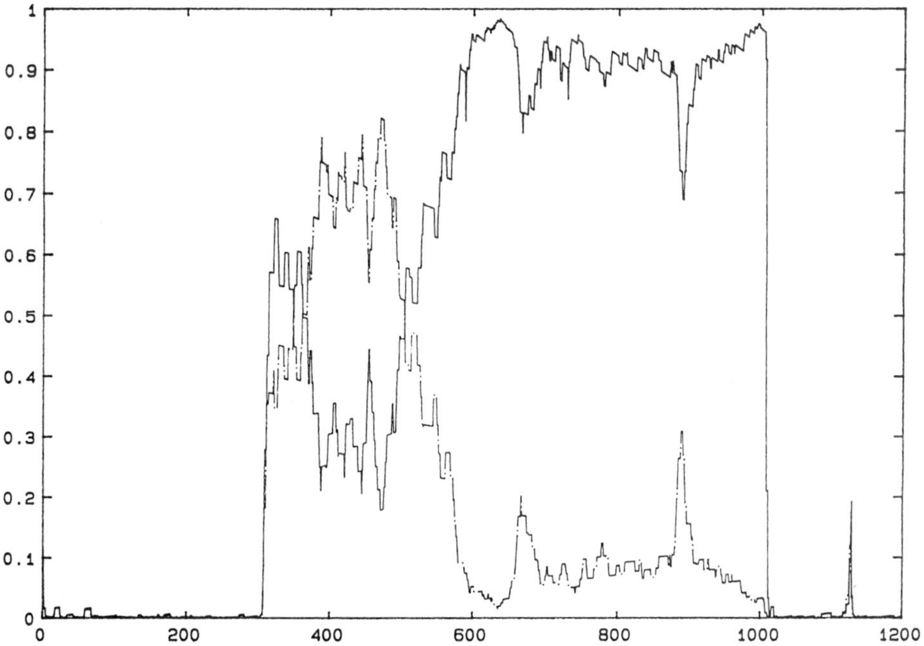

Figure 6.11 - Estimate trajectories with a high rate.
(— $\hat{\phi}_{t1}$, - - - $\hat{\phi}_{t3}$, $\hat{\phi}_{t2}$ not shown)

6.4 Filter performance

The filters obtained in § 6.2 and 6.3 result from the minimization of the mean square error and this is our basic rational for using them in practical situations. However additional analyzis of their performance is needed from the point of view of convergence and realistic implementations.

In the classical LQG setting the optimal filter (the Kalman filter) gives an unbiased estimate of the state and the estimator dynamics are stable. The filters presented in this chapter are non linear and the corresponding convergence and stability results are difficult to establish. A related analysis may be found in (Hijab, 1987) for the case where there are no jumps ($\Pi = 0$, unknown constant regime $r \in \{1, 2, ..., N\}$). The key notion is consistency : the filter is said *consistent* if $\lim_{t\to\infty} \hat{\phi}_t = \phi(r)$, that if the a posteriori probabilities vector converges to the actual regime indicator as more and more measurements become available. It is then shown (Hijab, 1987, theorem 4 p. 76) what conditions are to be imposed on the observation channel to guarantee consistency of the estimate. In the case of a direct regime measurement $dy_t = h(r)\,dt + dv_t$, a necessary and sufficient condition is $\int_0^\infty | h_j - h(r) |^2 dt = +\infty$ for $j \neq r$ (it is assumed that there is a non zero probability of being in any regime at $t = 0$). If the h_i, $i = 1$ to N, are moreover constant with time, the simplest condition is obtained as $h_i \neq h_j$ when $i \neq j$. This further explains the results of example 2 where the closeness of h_2 and h_3 confused the filter after the jump from the regime 1 to regime 3. In the no jump case it is interesting to relate consistency to the information contained in the measurements. For two distributions p_i and p_{0i}, $i = 1$ to N, over $S = \{1, 2, ..., N\}$ Shannon information is defined as $I(p, p_0) = \sum_{i=1}^{N} p_i \log \frac{p_i}{p_{0i}}$. It is easily seen that I is bounded ($0 \leq I(p, p_0) \leq - \sum_{i=1}^{N} p_{0i} \log p_{0i}$) and (Hijab, 1987) observed that I(t), defined as the information $I(\hat{\phi}_t, p_0)$, is simply given by

$$I(t) = E\,\{\frac{1}{2} \int_0^t | h(r) - \sum_{i=1}^{N} \hat{\phi}_{ti} h_i |^2 dt\,\}.$$

It follows that the filter is consistent if and only if the information I(t) attains its maximum at t goes to infinity ($I(\infty) = - \sum p_{0i} \log p_{0i}$). In the case where the regime jumps this theory does not apply without modifications. In particular consistency can then only be defined with respect to the ergodic distribution p_i^∞ (see chapter 2) and the h(r) term in the above expression of I(t) becomes time varying $h(r_t)$. However, as shown by example 2, this distance between h_i and h_j for $i \neq j$ still rules the richness of the observation signals and a good detection requires that the difference $h_i - h_j$, for $i \neq j$, is above the noise level. For $\Pi \neq 0$ the filter expects a transition and therefore it has a tendancy to see one in noise. As observed on the simulation this results in a bias and, even after a quiescent period, the probability of the most likely regime does not completely converge near 1.

The optimal filter representations given in previous sections produce the estimates as solutions of Ito's stochastic differential equations. This follows from a direct application of the general theorem of optimal filtering, as is recalled in appendix 1. Using this representation, we were able to conduct simulations (see examples of § 6.2 and 6.3 for, respectively, Wiener driven and point process observations) that demonstrated conclusively the qualitative properties of the estimators. It is not the purpose of this work to analyze in detail the implementation of an optimal filter on a digital computer and, in this respect, we would be content with the results of § 6.2 and 6.3. However, it is worth pausing for a while to provide some background on transformations associated with an efficient implementation. To do this we shall consider the simplest situation in § 6.2 whence observations are

$$dy_t = h(r_t)\, dt + dv_t$$

The optimal filter is recalled from theorem 6.1

$$d\hat{\phi}_{ti} = (\Pi'\hat{\phi}_t)_i\, dt + \hat{\phi}_{ti}\, (h_i - \sum_{j=1}^{N} \hat{\phi}_{tj}\, h_j)\, (dy_t - \sum_{j=1}^{N} \hat{\phi}_{tj}\, h_j\, dt)$$

$$i = 1 \text{ to } N$$

where the intensity of v_t has been normalized (V = 1). The i-th component of the estimate $\hat{\phi}_t = E\,\{\phi_t \mid Y_t\}$ is just the a posteriori probability $\hat{\phi}_{ti} = \mathcal{P}\,\{\phi_{ti} = 1 \mid Y_t\}$. Despite this appealing interpretation, it is known (Mitter, 1983) that working with the normalized estimate ($\sum_{i=1}^{N} \hat{\phi}_{ti} = 1$) is not practical because of the non linearities in the up-date term and the measurement level ($\sum_{j=1}^{N} \hat{\phi}_{tj} h_j$) estimates. Because of this modulation, it is only when the measurement noise is white that a direct physical interpretation of the filter equation is possible. As observed by (Clark, 1978) the necessity of this whiteness assumption is not satisfactory for real world applications where the spectral densities decrease for high frequencies. Replacing the above Ito representation by its Stratonovitch equivalent would not alleviate this difficulty and another representation is needed to avoid the modulation phenomena. This is provided by Zakai's equation (1969), which is another representation of the optimal filter in terms of unnormalized estimates. For the above model, Zakai's equation reads

$$d\breve{\phi}_{ti}) = (\Pi'\breve{\phi}_t)_i \, dt + \breve{\phi}_{ti} \, h_i \, dy_t$$
$$i = 1$$

and the normalized estimates are recovered by $\hat{\phi}_{ti} = \breve{\phi}_{ti} / \sum_{j=1}^{N} \breve{\phi}_{tj}$, $i = 1$ to N. Using $\breve{\phi}_t$ instead of $\hat{\phi}_t$ we get rid of the non linearity but a modulation of dy_t by $\breve{\phi}_t$ remains . This can be eliminated by considering the logarithm of $\breve{\phi}_t$. For $\breve{\phi}_{ti} > 0$, $i = 1$ to N, applying the differentiation rule gives

$$d\ln\breve{\phi}_{ti} = \frac{1}{\breve{\phi}_{ti}} (\Pi'\breve{\phi}_t)_i \, dt + h_i \, dy_t - \frac{1}{2} h_i^2 \, dt$$

which can be interpreted as a Wiener differential fitting in a real world situation. A last representation of the optimal filter might be useful in connection with the change detection problem outlined in §.6.1. To test the hypothesis "$r_t = i$" against the hypothesis "$r_t = i_0$" based on the information accumulated in Y_t the normalized estimates are not needed since obviously $\breve{\phi}_{ti} / \breve{\phi}_{ti_0} = \hat{\phi}_{ti} / \hat{\phi}_{ti_0}$. Indeed the likelihood ratio of hypothesis "$r_t = i$" over "$r_t = i_0$", noted here $\mathscr{L}_t(i, i_0)$, can be expressed in terms of $\breve{\phi}_{ti} / \breve{\phi}_{ti_0}$ and the unconditional a priori probabilities p_{ti} and p_{ti_0} ($dp_{ti} = (\Pi' p_t)_i \, dt$, $i = 1$ to N) as

$$\mathscr{L}_t(i, i_0) = \frac{p_{ti_0}\, \breve{\phi}_{ti}}{p_{ti}\, \breve{\phi}_{ti_0}}$$

Propagating the $\breve{\phi}_{ti}$'s, i = 1 to N, in their logarithmic form (together of course with the ordinary differential equations for the p_{ti}s) is therefore sufficient for regime estimation and change detection and in practice the normalized estimate equation would often not be propagated.

Discretization of the continuous-time filters for implementation on a computer is not a trivial task. One difficulty is the choice of the sampling period and integration method, with an obvious compromise between the precision of the approximation and the computer time. Compared to the corresponding questions for ordinary differential equations, little is known about this for stochastic differential equations. Several types of convergence may be considered, either in terms of moments, e.g. a bound on the distance between the approximated mean square and its true value, or in terms of sample paths, e.g. a bound on the distance between an approximated trajectory and the true trajectory. Some theoretical results are known regarding comparison of integration schemes of varying complexity, from Euler schemes to Runge-Kutta schemes of high order. References (Wright, 1974, Rümelin, 1982, Pardoux and Talay, 1985, Liske and Platten, 1987) and the thesis (Newton, 1983) provide an introduction to the subject but work remains to be done on the conception of efficient numerical schemes for the simulation of stochastic differential equations. For the examples of § 6.2 and 6.3 the objective was to obtain representative trajectories of the optimal filter without special requirements on numerical accuracy and we used the simplest schemes (Euler) with an adjusted (constant) sampling period.

6.5 Notes and References

Using the powerful tools of optimal filtering theory, various estimators have been proposed to estimate unmeasured regime or/and state variables. Depending on the specificities of a given application, Wiener driven or point process observations, or a combination, give the measurement model. The material presented follows directly from the general theory, see for example (Wong and Hajek, 1985, chapter 7) or the summary given in appendix 1. There are, however, many important related questions concerning optimal filtering, and the references of § 6.4 provide roots to probe further into this rich subject.

The filter in theorem 6.1 is sometimes called Wonham's filter from the pioneering paper (Wonham, 1964). Since then it has been rederived many times and adapted to a richer class of observations as in theorems 6.2 and 6.3. When the state is not available the optimal filter is infinite dimensional as discussed from theorem 6.4 and some approximations have to be made. Working in discrete-time, where the dimensionality problem translates into the explosively growing number of histories that must be stored, Blom recently proposed an attractive suboptimal filter, called the Interacting Multiple Model (IMM, Blom 1984). It is worth noting that Blom's work was precisely motivated by the tracking problem with sudden accelerations discussed in chapter 1. The interest of this approximation is that it provides a precision comparable with higher order solutions (such as the so-called Generalized Pseudo Bayes algorithms). Indeed, as with Kalman filtering, we sometimes prefer an observer, that is, a suboptimal filter reproducing the optimal structure but with a non optimal gain.

There exists classical methods for the testing of changes in signals and systems, and most prominently, the Generalized Likelihood Ratio (GLR) and Multiple Model (MM) tests. Even though they are concerned with estimating the value of an unknown constant regime these techniques have some relevance to our problem. In particular, when regime transitions are slow (small entries for the Π matrix) a reasonable approximation treats the regime as a constant unknown variable and GLR or MM methods can be used. In the basic GLR algorithm, regimes are identified by means of their signature, that is the correlations in the innovation process that are due to additive regime driven signals. These signatures are generated by post-processing the output of a Kalman filter tuned to the nominal regime parameters. In the MM algorithm, a bank of Kalman filters are run in parallel and the true regime is signalled by the whiteness of its innovation process. The survey (Willsky, 1976) reviews these algorithms and related variants. An important assumption for GLR and MM solutions was the knowledge of either a nominal model or a list of models, but this has been relaxed for the GLR algorithm (Basseville 1986) by estimating on-line not only the change but also some nominal model parameters. For an up-date on this domain, the conference proceedings (Basseville and Benveniste,

1986) are recommended with recent results on the approximation and convergence problems.

On the material in § 6.3 (Brémaud, 1981) is our main reference. With queuing applications in mind, it provides the required background on point process dynamics and solves several control and filtering problems beyond the limited illustration given by theorem 6.4. For a treatment of these topics without emphasis on martingale calculus, the classical book (Snyder, 1975) or papers such (Rudemo, 1972) may be consulted. The martingale approach is exemplified by (Segall, 1973) and presented in details in Brémaud's monograph.

This chapter provides us with means for estimating regime transitions when they are not observed. The reconfiguration of the control law for a hybrid system requires a good regime estimation and § 6.2 and 6.3 show that this can indeed be accomplished. However, simulations also reveal that during short transients there are significant discrepancies between $\hat{\phi}_t$ and ϕ_t, when false alarms occur or when detection is delayed (see figure 6.3). To solve the global control problem, the output of the regime filters (Theorems 6.1, 6.2, 6.3 or 6.4) is fed into the plant state dynamics through a reconfiguration, or more generally an adaptation, of the control law. An accurate design of this adaptive law must therefore take into account the errors of the regime detection filter. This is the purpose of chapter 7.

References

Basseville, M., and Benveniste, A., Eds (1986). Detection of Abrupt Changes in Signals and Dynamics Systems, Lecture Notes on Control and Information Sciences, Springer-Verlag, Berlin.

Basseville, M. (1986). Détection de pannes et reconfiguration automatique, Proc. Colloque l'Automatique pour l'Aéronautique, Paris, pp. 431-453.

Blackwell, D., and Girschick, M.A. (1954). Theory of Games and Statistical Decisions, J. Wiley, New-York.

Blom, H.A.P. (1984). An efficient filter for abruptly changing systems, Proc. 23rd IEEE Conf. Decision Control, Las Vegas, pp. 656-658.

Brémaud, P. (1981). Point Processes and Queues - Martingale Dynamics, Springer-Verlag, New-York.

Clark, J.M.C. (1978). Panel discussion "Are Ito Calculus and martingale theory useful in practice ?" in Communication Systems and Random Process Theory (J.K. Skwirzinski, ed.) NATO Adv. Study Inst., Sijthoff and Noordhoff, Amsterdam, p. 753.

Davies, M.H.A. (1975). The application of nonlinear filtering to fault detection in linear systems, IEEE Trans. Aut. Control, AC-20 : 257.

Ferguson, T.S. (1967). Mathematical Statistics a Decision Theoretic Approach, Academic Press, New-York.

Hijab, O. (1987). Stabilization of Control Systems, Springer-Verlag, Berlin.

Liske, H., and Platen, E. (1987). Simulation studies on time discrete diffusion approximations, Math. Comp. Sim., 29 : 253.

Mitter, S.K. (1983). Approximations for non linear filtering, in Non Linear Stochastic Problems (R.S. Bucy and E.F. Moura, eds), Reider Pub. Co., Dordrecht.

Newton, N.J. (1983). Discrete approximations for markov chain filters, PhD dissertation, Imperial College, London.

Pardoux, E., and Talay, D. (1985). Discretization and simulation of stochastic differential equations, Acta Appl. Math. , 3 : 23.

Rümelin, W. (1982). Numerical treatment of stochastic differential equations, SIAM J. Numer. Anal. , 19 : 604.

Rudemo, M. (1972).Doubly stochastic poisson processes and process control, Adv. Appl. Prob. , 4 : 318.

Segall, A. (1973). A martingale approach to modelling, estimation and detection of jump processes, PhD Dissertation, Stanford University.

Snyder, D.L. (1975). Random Point Processes, J. Wiley, New-York.

Willsky, A.S. (1976). A survey of design methods for failure detection in dynamic systems, Automatica, 12 : 601.

Wong, E., and Hajek, B. (1985). Stochastic Processes in Engineering Sciences, Springer Verlag, New-York.

Wonham, W.M. (1964). Some applications of stochastic differential equations to optimal non linear filtering, Tech. Report 64-3, RIAS, Baltimore, Md.

Wright, D.J. (1974). The digital simulation of stochastic differential equations, IEEE Trans. Aut. Control , AC-19 : 75.

Zakai, M. (1969). On the optimal filtering of diffusion processes, Z. Wahrscheinlichkeitstheorie verw. Geb. , 11 : 230.

7

Control Under Regime Uncertainty

7.1 Jump Detection and Dual Control

The filters of chapter 6 process sensor data to estimate the current regime probabilities. Associated with a detection logic, as explained in § 6.1, they deliver the most plausible value of the regime at any instant. This decision can be exploited for the purpose of controlling the state of the hybrid model and the simplest solution is to use the output of the decision test exactly as if it were the true regime. This is a certainty equivalence solution, in the sense discussed in chapter 5. The regulator is then designed with any of the algorithms of chapters 3, 4 or 5 and implemented by using the decided value in place of r_t.

The deficiency of the certainty equivalence approach is that it ignores the errors of the filter and detector that will inevitably corrupt the decided variable. In some situations this regime uncertainty significantly degrades performance and must be accounted for in the regulator design model. The purpose of this chapter is to propose two regulators dealing with regime uncertainty. The first one (§7.2) is suited to applications where sensors do not separate regimes beside a rough clustering, leading to small but persistent errors within a cluster. In § 7.5 we consider the converse situation when detection delays and false alarms lead to large but transient regime misclassification.

The basic idea is to capture the uncertainty in a simple model, static in the first case (§ 7.2) and dynamic in the second one (§ 7.3), that can be included in the control design model. The dynamic model of § 7.3 for delays and false alarms is also used in § 7.4 to show that a certainty equivalence reconfiguration of the regulator gain sometimes even destabilizes the system.

The control action is not decoupled from the regime estimation filter. It excites the state dynamics and thus influences the estimate trajectories, see for example the filter of theorem 6.3 where the x_t dynamics would be driven by a control signal. It is intuitive that some control signals are more favourable than others for regime estimation and that the regulation objective itself is detrimental to the use of an x_t measurement in the filter : as mentionned about example 2 in § 6.2, if the regulator is very successful x_t stays close to zero and the sensor sees only noise. The contradiction between the regulation and the excitation contents of the control signal is captured by Feldbaum's dual control theory, the purpose of which is to resolve this trade-off through the optimization of a global performance index (Feld'baum 1965). This theory is however difficult and only suboptimal solutions have been obtained so far, more precisely heuristic solutions that tend to preserve the qualitative properties of the optimal and dual solution.

The approach used in the sequel is to constrain the class of admissible control laws so that dual solutions are excluded. In this smaller class the optimization problem becomes tractable and complete design equations are given. We shall discuss again the implications of the dual control in § 7.6 together with a few ideas on the performance improvement that could be achieved with the dual optimal regulator.

7.2 Influence of a Small Regime Error

The jump detection scheme discussed in chapter 6 infers the (a posteriori) most likely regime ℓ_t from the mean square estimate $\hat{\phi}_t$. It is therefore the dynamics of $\hat{\phi}_t$ which ultimately characterize the imperfections of the detected variable and the purpose of § 7.3 is to propose a black-box model for this part of the overall control system. However, at an intermediate level of complexity, it might be worth in some applications to describe ℓ_t errors in a simpler way, for example in terms of static misclassification probabilities. Although this is a rather coarse description and does not fully exploit the filter data, it will become clear that the performance can be quite satisfactory if the classification error is small.

To illustrate this design philosophy, the solar thermal receiver of chapter 1 is considered again. In this application the regime variable is a quantized version of the solar flux across the receiver panel and six cells, placed at various points on the panel, are used to sense it. Because underestimating the insolation would lead to an insufficient feedwater flow rate and might cause structural damage to the fragile panel, it was decided at an early stage to select the largest among the six sensor readings as the insolation value used in the feedwater regulation loop. This way of ensuring some safety margin introduces a bias towards overestimated insolation values, however it can be expected that the bias, though not negligible, is small. Typically the "sensed" insolation, meaning the value computed by the "high select" logic, is within 20% of the true insolation: if $\varphi(r_t)$ is the insolation when the regime is r_t and if ℓ_t is the current value of the perceived regime, the small uncertainty introduced by sensor errors can be described, for example, by $\mathcal{P}\{\varphi(\ell_t) = \varphi(r_t)\} = 0.5$ and $\mathcal{P}\{\varphi(\ell_t) = 1.2\,\varphi(r_t)\} = 0.5$, reflecting that the sensed insolation is, on the average, exact half of the time and has a 20 % positive bias the rest of the time.

Based on this discussion, the model of the regime is temporarily modified (we shall go back to our basic model in § 7.3). The new model is as follows :

For $r_t \in S$ $(S \in \{1, 2, \dots, N\})$ consider $T_1, T_2, \dots, T_{N0}$ a partition of S. The perceived operating conditions are indexed by the projection ℓ_t, $\ell_t \in S^0 = \{1, 2, \dots, N^0\}$ with $N_0 \le N$ and it is assumed that the dynamics of ℓ_t are given by a Markov chain

$$d\phi_t = \Pi^{0'}\phi_t\,dt + d\mu_t \tag{7.2.1}$$

where ϕ stands for the ℓ_t indicator and μ_t is a Ξ-martingale as before. Transitions of the true regime are now in concert with that of ℓ_t and the distribution of r_t, assumed independent on the past of ℓ_t, is given by

$$\mathcal{P}\{r_t = j \mid \ell_t = i\} = \begin{cases} d_{ji} & \text{if } j \in T_i \\ 0 & \text{otherwise} \end{cases} \tag{7.2.2}$$

in other words ℓ_t perceives the correct subset of S to which r_t belongs but it is unable to distinguish the true value in the subset. The a priori distribution is the only way we have to distinguish regimes within a class T_i. The small error assumption manifests itself through the fact that all the regimes in the same subset T_i are close (for example with the solar receiver one could have two values of ℓ_t, $\ell_t = 1$ cloudy, $\ell_t = 2$ clear and $S = \{1, 2, 3,4\}$, $T_1 = \{1, 3\}$, $T_2 = \{2, 4\}$ and $\varphi(3) = 1.2\varphi(1)$, $\varphi(4) = 1.2\,\varphi(2)$). This situation corresponds to a cluster of regimes when the sensors are not able to separate regimes that are close to one another.

Of course the uncertainty on the regime impacts the perceived dynamics of the plant state and this should be reflected by the regulation laws. For reference, recall the state dynamics

$$\dot{x}_{pt} = f(x_{pt}, u_{pt}, r_t)$$

The purpose is to regulate x_{pt} around some nominal set-point x_n while using limited excursions of u_{pt} from the corresponding nominal control u_n. Usually, the operating point is picked from the plant equilibria

$$0 = f(x_n, u_n, r) \tag{7.2.3}$$

using considerations like performance, safety, ... which are not elaborated here. As (7.2.3) points out the nominal values are regime dependent ($x_n = x_n(r)$, $u_n = u_n(r)$), like for example with the solar receiver when higher insolation calls for higher nominal feedwater flow rate. This simple fact has far reaching consequences in the presence of regime uncertainty because it implies that the regulator does not know for sure the set-points around which x_{pt} and u_{pt} are to be stabilized.

The difficulty here is that r_t is not known besides $r_t \in T_i$ when $\ell_t = i$ and a static measure of this uncertainty is captured by the d_{ji} in (7.2.2). The error variables therefore have to be defined with respect to $x_n(\ell_t)$ and $u_n(\ell_t)$ instead of $x_n(r_t)$ and $u_n(r_t)$ as would be the case if the regime were observed

$$\begin{aligned} x_t &= x_{pt} - x_n(\ell_t) \\ u_t &= u_{pt} - u_n(\ell_t) \end{aligned} \tag{7.2.4}$$

It is hoped that the regulation errors remain small most of the time and it is thus legitimate to describe the dynamics of x_t as a linearized approximation of the non linear x_{pt} dynamics. Define

$$\begin{aligned} A_i &= f_{x_p}(x_n(i), u_n(i), i) \\ B_i &= f_{u_p}(x_n(i), u_n(i), i) \end{aligned}$$

where f_{x_p} and f_{u_p} stand for the partial derivatives with respect to x_p and u_p (usual smoothness conditions on f are assumed, see chapter 1). Then for small x_t and u_t one has approximately

$$\dot{x}_t = A_i x_t + B_i u_t + A_i (x_n(j) - x_n(i)) + B_i (u_n(j) - u_n(i)) \tag{7.2.5}$$

when $\ell_t = j$, $r_t = i$

Due to the presence of r_t, the regulator is not able to use this model directly and it has to average over the residual uncertainty (7.2.2). Of course this averaging is legitimate only when the A_is and B_is do not vary too much for values of r_t belonging to the same class. An additional peculiarity occurs at instants when the regime jumps : because x_{pt}, the physical variable, can reasonably be assumed continuous, (7.2.4) reveals that the state deviation experiences a discontinuity, say when ℓ_t goes from $\ell_{t-} = i$ to $\ell_t = j$, $i, j = 1$ to N^0,

$$x_t = x_{t-} + x_n(i) - x_n(j) \tag{7.2.6}$$

The design model is now complete, (7.2.2) describing the detection errors, (7.2.5) the error dynamics between jumps and (7.2.6) the discontinuity at jump instants. We define $\delta(i, j) = x_n(j) - x_n(i)$ and $\Delta(i, j) = A_j\,\delta(j, i) + B_j\,(u_n(i) - u_n(j))$ to simplify notations.

The observations available at the regulator are the state (it is assumed here that it is possible to measure the plant state) and the perceived regime ℓ_t. With $X_t = \sigma - \{x_s, t_0 \leq s \leq t\}$ and $L_t = \sigma - \{\ell_s, t_0 \leq s \leq t\}$ the corresponding σ-algebras, the class $\mathcal{U}$ of admissible control policies consists of $X_t \vee L_t$- measurable functions satisfying familiar smoothness conditions (see appendix 2) and the control design is formulated as an optimization problem

$$\underset{u \in \mathcal{U}}{\text{Min}}\ J = E\{\int_{t_0}^{t_f} (x'Q_1x + u'Q_2u)\,dt \mid x_{t0}, \ell_{t0}, t_0\} \tag{7.2.7}$$

However the optimal control is then of the dual type discussed in § 7.1, in the sense that the regulator will attempt to learn the value of the regime based on $X_t \vee L_t$. To avoid the associated difficulties, we decide to further restrict the class of admissible policies to memoryless solutions, that is instantaneous feedback of (x_t, ℓ_t). It can be intuitively asserted that if ℓ_t and r_t are close the so constrained optimal regulator performs almost like its unconstrained counterpart, since then learning would any way provide only a marginal reduction of the uncertainty. Using dynamic programming, this is translated in the form of the cost-to-go V. It should be parameterized as $V = V(x_t, \hat{\phi}_t, t)$, where $\hat{\phi}_t$

where the scalars μ_i, i = 1, N^0, satisfy

$$
\begin{aligned}
-\dot{\mu}_i = & \sum_{j\in T_i} d_{ji}\, \delta(j, i)'\, Q_1\, \delta(j, i) - \lambda_i' \Big(\sum_{j\in T_i} d_{ji}\, B_j\Big)'\, Q_2^{-1} \Big(\sum_{j\in T_i} d_{ji}\, B_j\Big) \lambda_i \\
& + 2 \Big(\sum_{k=1}^{N^0} d_{ji}\, \Delta(i, j)\Big)'\, \lambda_i \qquad (7.2.12) \\
& + \sum_{k=1}^{N^0} \pi_{ik}^0 \left(\mu_k + 2\lambda_k'\, \delta(k, i)' + \delta(k, i)'\, \lambda_k\, \delta(k, i)\right)
\end{aligned}
$$

$$\mu_i(t_f) = 0$$

Proof :

The cost-to-go is defined as

$$\operatorname*{Min}_{u\in \mathcal{U}}\ E\{ \int_t^{t_f} (x'Q_1x + u'Q_2u)\, dt \mid X_t \vee L_t\}$$

and, by assumption (7.2.8), we parameterize it as $V(x_t, \ell_t, t)$. The Hamilton-Jacobi-Bellman equation is then

$$
\begin{aligned}
-V_t(x, \ell, t) = & \operatorname*{Min}_{u\in \mathcal{U}} E\{(x + \delta(r, \ell))'\, Q_1\, (x + \delta(r, \ell)) + u_t'\, Q_2\, u_t \\
& + \mathcal{L}_u V(x, \ell, t) \mid x, \ell, t\} \qquad (7.2.13)
\end{aligned}
$$

where $\mathcal{L}_u$ is the generator as in chapter 3

$$\mathcal{L}_u V(x, i, t) = (A(r_t)\, x + B(r_t)\, u_t + \Delta(i, r_t))'\, V_x(x, i, t) + \sum_{k=1}^{N^0} \pi_{ik}^0\, V(x, k, t)$$

To compute the minimizing control, we bring together the terms involving u_t

$$E\, \{u_t'\, Q_2\, u_t + (B(r_t)\, u_t)'\, V_x(x, i, t) \mid x, i, t\}$$

Averaging over r_t this gives

$$u_t'\, Q_2\, u_t + u_t' \Big(\sum_{j\in T_i} d_{ji}\, B_j\Big)'\, V_x(x, i, t)$$

stands as in chapter 6 for the estimate E $\{\hat{\phi}_t \mid X_t \vee L_t\}$ with $\phi = \phi(r) \in \mathbf{R}^N$ the true regime indicator, but the above simplification replaces this parameterization by

$$V(x_t, \ell_t, t) \tag{7.2.8}$$

The solution is then found by expanding V and averaging over the r_t uncertainty using the distribution (7.2.2). The result is

<u>Theorem 7.1</u>

The solution of the optimization problem (7.2.7), for the hybrid model (7.2.5), (7.2.1), (7.2.2) with assumption (7.2.8) is

$$u_t^* = -Q_2^{-1}\Big(\sum_{j\in T_i} d_{ji} B_j\Big)' (\Lambda_i x_t + \lambda_i) \qquad \text{when } \ell_t = i$$

where the matrices Λ_i and vectors λ_i, $i = 1, N^0$, satisfy

$$-\dot{\Lambda}_i = \Big(\sum_{j\in T_i} d_{ji} A_j\Big)' \Lambda_i + \Lambda_i \Big(\sum_{j\in T_i} d_{ji} A_j\Big) - \Lambda_i \Big(\sum_{j\in T_i} d_{ji} B_j\Big)' Q_2^{-1} \Big(\sum_{j\in T_i} d_{ji} B_j\Big) \Lambda_i$$
$$+ Q_1 + \sum_{k=1}^{N^0} \pi_{ik}^0 \Lambda_k \tag{7.2.9}$$

$$\Lambda_i(t_f) = 0$$

and

$$-\dot{\lambda}_i = \Big(\sum_{j\in T_i} d_{ji} A_j - \Big(\sum_{j\in T_i} d_{ji} B_j\Big) Q_2^{-1} \Big(\sum_{j\in T_i} d_{ji} B_j\Big)' \Lambda_i\Big) \lambda_i$$
$$+ Q_1 \sum_{j\in T_i} d_{ji}\, \delta(j, i) + \Lambda_i \sum_{j\in T_i} d_{ji}\, \Delta(i, j) + \sum_{k=1}^{N^0} \pi_{ik}^0 (\lambda_k + \Lambda_k\, \delta(k, i))$$
$$\tag{7.2.10}$$

$$\lambda_i(t_f) = 0$$

The corresponding cost is

$$J = x_{t0} \Lambda_{i0}(t_0)\, x_{t0} + 2\lambda_{i0}(t_0)' x_{t0} + \mu_{t0}(t_0) \qquad \text{when } \ell_{t0} = i_0 \tag{7.2.11}$$

and the optimal control

$$u_t = -\frac{1}{2} Q_2^{-1} (\sum_{j \in T_i} d_{ji} B_j)' V_x (x, i, t) \qquad \text{when } \ell_t = i \tag{7.2.14}$$

Bringing this expression back into (7.2.13) we obtain a partial differential equation in V. Let us guess a solution of the form

$$V(x, i, t) = x' \Lambda_i x + 2\lambda_i' x + \mu_t$$

With this, (7.2.14) immediately gives u_t^* and (7.2.13) reduces, at the minimum, to a quadratic form in x with left hand side

$$V_t (x, i, t) = x' \dot{\Lambda}_i x + 2\dot{\lambda}_i' x + \dot{\mu}_i \qquad \text{when } \ell_t = i$$

The coefficients of the quadratic, linear and constant terms in x, averaging on r_t wherever needed, give, with some lengthy but straightforward computations, (7.2.9), (7.2.10) and (7.2.12) respectively. Boundary conditions $\Lambda_i(t_f) = 0$, $\lambda_i(t_f) = 0$ and $\mu_i (t_f) = 0$, i = 1 to N^0, result from the definition of the cost-to-go at $t = t_f$. So does (7.2.11) at $t = t_0$, and the proof is completed. □

The control law obtained has a familiar form with a linear state deviation feedback and a bias. The feedback gain $-Q_2^{-1} (\sum_{j \in T_i} d_{ji} B_j)'\Lambda_i$ is precisely the gain for an averaged system where $A(r_t)$ and $B(r_t)$ would be replaced by their mean value conditioned on ℓ_t. We note that the weighting matrices Q_1 and Q_2 could be dependent on ℓ_t as well without modifying the solution ($Q_1 = Q_{1i}$ when $\ell_t = i$ ...). The bias term depends on the state deviation model forcing function $\Delta(i, j)$ and it also reflects the occurence of jumps by displacing the steady-state away from the current $x_n(i)$ towards future $x_n(k)$, k = 1, N^0 such that $\pi_{ik}^0 \neq 0$. It can be expected that the regulator given by theorem 7.1 will tend to be less sensitive to modeling errors thanks to the way it incorporates regime uncertainty.

The main limitation of theorem 7.1 lies with the "cluster of regimes" assumption and when the regime uncertainty is large other phenomena must be taken into account. This is the purpose of §7.4 and 7.5, based on a model introduced below.

7.3 A Model of the Regime Estimation Filter

The controller presented in the previous section takes into account the uncertainty surrounding the current regime of operation through the d_{ji}, $i = 1, N^0$, $j = 1, N$, a priori probabilities. However it does not attempt to reduce this uncertainty as more information becomes available, in other words it does not include any learning capacity. It is apparent that the regime estimation filters of chapter 6 would be useful in this respect since they extract, in a mean square optimal sense, the information on r_t contained in the observation O_t. To exploit these results, we abandon the (ℓ_t, r_t) model of § 7.2 and return to our basic Markov chain model for r_t transitions (see chapter 1).

The output of any of these filters has rather complicated dynamics (see for example figure 6.3 or figure 6.5). Our objective now is to mix this with a regulation law and, to arrive at a tractable model, we need to simplify the $\hat{\phi}_t$ behaviour by retaining only its dominant qualitative characteristics. Recall from chapter 6 that a decision is taken based on $\hat{\phi}_t$, picking up the most plausible regime, noted ℓ_t. Even though we use the same notation for simplicity there should be no confusion with the ℓ_t variable of § 7.2. Both quantities are related to our perception of the current regime, but it is now an estimate while in § 7.2 it was the observed projection of r_t on to S^0. Looking at, say, figure 6.7, we see that it is rather easy to select a threshold such that ℓ_t tracks r_t very well. However it is not possible to suppress false alarms and detection delays (look at figure 6.3 for $t \sim 130$, $t \sim 300$, $t \sim 850$ or $t \sim 1000$). Around these instants there is a significant discrepancy between ℓ_t and r_t, corresponding to a wrong decision, say $\ell_t = 1$ when $r_t = 2$. The technique of § 7.2 is well suited for the case where there is a small uncertainty on the regime : typically clusters of regimes such that the sensors are not able to separate the regimes within a cluster but give only the cluster to which the current regime belongs.

The situation we have in mind here is quite different, with large and infrequent errors, corresponding to false alarms and delays in a failure detection and identification system. Of course both kinds of errors might be present but we shall distinguish the two cases for the benefit of clarity.

The two deficiencies of ℓ_t that we want to discuss now are false alarms and detection delays. The discrete variable $\ell_t \in \{1, 2, \dots, N\}$ is thus of the same nature as the true regime and indeed its evolution is driven by the evolution of r_t reflected in $\hat{\phi}_t$. It is therefore natural to define ℓ_t conditionally on r_t. More precisely, we shall assume that the following model describes the detection delay :

When r_t has jumped from i to j, ℓ_t follows with a delay τ that is an independent exponentially distributed random variable with mean $1/\pi^0_{ij}$. This is written as

$$\mathcal{P}\{\ell_{t+dt} = j \mid \ell_s = i,\ s \in [t_0, t],\ r_{t_0} = j,\ r_{t_0^-} = i\} =$$

$$\begin{cases} \pi^0_{ij}dt + o(dt) & i \neq j \\ 1 + \pi^0_{ii}dt + o(dt) & i = j \end{cases} \tag{7.3.1}$$

The entries of the matrix $\Pi^0 = (\pi^0_{ij})_{i,j=1,N}$ are evaluated from observed sample paths, as shown in the example of § 7.5 below. False alarms are described in similar terms with occasional declaration of an ℓ_t transition from i to j when r_t in fact remained at i. An independent exponential distribution with rate π^1_{ij} is again assumed

$$\mathcal{P}\{\ell_{t+dt} = j \mid \ell_s = i,\ s \in [t_0, t]\} =$$

$$\begin{cases} \pi^1_{ij}dt + o(dt) & i \neq j \\ 1 + \pi^1_{ii}dt + o(dt) & i = j \end{cases} \tag{7.3.2}$$

with a matrix $\Pi^1 = (\pi^1_{ij})_{i,j=1,N}$ of false alarm rates.

The merit of the model (7.3.1), (7.3.2) for ℓ_t dynamics is that it captures in a quantitative way (the data being the matrices Π^0 and Π^1) the imperfections of the detection algorithm. It is the simplest model with this property and this simplicity pays off in § 7.4 and 7.5 when it allows the derivation of explicit analysis and synthesis results. It is not claimed however that this is a very accurate description, especially the exponential distribution hypothesis is certainly questionable for applications where false alarms tend to occur in clusters. Refinements of (7.3.1) and (7.3.2) using a larger chain for ℓ_t and non markovian conditional jumps could be envisionned (see the non markovian transitions of chapter 8) but this is not pursued here.

Although it was introduced from the filters discussed in chapter 6, we believe that the model (7.3.1), (7.3.2) can serve as a black box representation for most of the rupture detection algorithms available in the literature, regardless of the technique actually used to generate a decision. The figures of merit of any detection device are its detection delays and false alarm rates and they are explicitly pictured in our model.

The interaction between detection and reconfiguration in a hybrid control situation is displayed on figure 7.1

The decision on the current regime is fed into the reconfiguration device to adapt the guidance and stabilization control laws. This is typically the case with the fault tolerance application of chapter 1. In signal processing applications (e.g. seismic data or EEG processing) the output of the decision device is the end result while here it is an intermediate result used to infer modifications of controlled dynamics. Obviously our situation is more delicate because we have to take additional care for the wrong decisions that severely impacts the reconfigured dynamics.

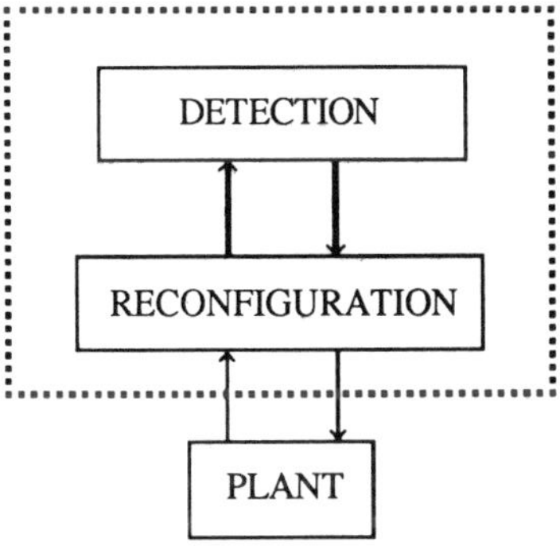

Figure 7.1 - The coupling between detection and reconfiguration.

We shall proceed in two steps. The stability of the closed-loop dynamics (the stabilization loop in figure 7.1) is first discussed when reconfiguration is based on ℓ_t instead of r_t as we had done previously (chapters 3 and 5). A synthesis for a mixed guidance and stabilization control law is proposed next, reflecting in the design model the imperfections of the detection and isolation algorithm.

7.4 Stability Analysis

There are situations where large errors on the regime have to be described. For example in the target tracking application of chapter 1 it is desirable to include in the model the case where the image-based portion of the system misclassifies a tank as a jeep, may be because of partial obscuration by a cloud of smoke.

Here we would like to analyze the influence of these large, but transient, errors on the stability of the plant with a reconfigured feedback. We use the model of § 7.3 for detection errors and again linearized dynamics for the state and control deviations $x_t = x_{pt} - x_n(i)$, $u_t = u_{pt} - u_n(i)$. However special attention must be paid here to the linearization procedure because the non-linearity is indexed by the actual regime r_t while the regulator knows ℓ_t, the output of the detection device (but ignores r_t). Following a transition in r_t, say from i to j, the regulator will thus wrongly try to maintain x_{pt} close to $x_n(i)$ until ℓ_t switches to j: linearization of the jth non-linearity should then be performed around the ith nominal. Two sets of matrices are therefore introduced

$$A(\ell, r) = f_{x_p}(x_n(\ell), u_n(\ell), r)$$

$$B(\ell, r) = f_{u_p}(x_n(\ell), u_n(\ell), r)$$

and we consider the linearized dynamics

$$\dot{x}_t = A_{ij} x_t + B_{ij} u_t \qquad \text{when } \ell_t = i \text{ and } r_t = j \tag{7.4.1}$$

where the regimes, estimated and actual, have been denoted by indices.

Though it will suffice here for closed-loop stability analysis, the model (7.4.1) will have to be refined in § 7.5 when we turn to the synthesis problem. This hybrid model with a double index also arises naturally in the target tracking application of chapter 1 (see (Blom, 1986)).

We restrict attention to linear state feedback indexed by the regime perceived at the regulator

$$u_t = \Gamma_i \, x_t \quad \text{when } \ell_t = i \tag{7.4.2}$$

so that the reconfiguration of the stabilization law consists in a switch of its gain based on the imperfect information delivered by the detection device.

The problem can be summarized as follows : how do errors on the regime, as described by (7.3.1) and (7.3.2), affect the stability of (7.4.1) when they are fed back in the control loop through (7.4.2) ? In addition we hope to receive a quantitative answer in terms of the detection delay and false alarm generators (Π^0 and Π^1).

Building on the results of chapter 2, we deal with exponential stability in the mean square sense (ESMS) and we obtain a sufficient stability condition that also guarantees almost sure stability.

Theorem 7.2

The system (7.4.1) with the control law (7.4.2) and regime errors (7.3.1), (7.3.2) is ESMS if, for positive definite matrices Q_{ij} , i, j = 1, N, the solutions P_{ij}, i, j = 1, N of the set of N^2 coupled Liapunov equations

$$0 = \tilde{A}_{ii}' \, P_{ii} + P_{ii}\tilde{A}_{ii} + Q_{ii} + \sum_{j=1}^{N} \pi_{ij} \, P_{ij} \; + \sum_{j=1}^{N} \pi_{ij}^{1} \, P_{ji}$$

$$i, j = 1, N \tag{7.4.3}$$

$$0 = \tilde{A}_{ij}' \, P_{ij} + P_{ij} \, \tilde{A}_{ij} + Q_{ij} + \pi_{ij}^{0} \, (P_{jj} - P_{ij})$$

with $\tilde{A}_{ij} = A_{ij} + B_{ij} \;\; \Gamma_i$

are positive definite.

Proof :

The proof is based on a result of (Kats and Krasovskii, 1960) adapted to a hybrid setting as in chapter 2 and we refer to the conditions given there.

With the P_{ij}, i, j = 1, N defined by (7.4.3) the quadratic function

$$V(x, \ell, r, t) = x'P_{ij}x \qquad \text{when } \ell_t = i \text{ and } r_t = j$$

is proposed as a candidate Liapunov function. It obviously satisfies conditions (i), (ii) and (iii) of theorem 2.3, with a double indexing (for fixed ℓ and r, it is continuous and has bounded continuous derivatives in x and t, etc). To compute $\mathscr{L}V$ and check condition (iv) we shall distinguish between $\ell = r$ and $\ell \neq r$ situations.

Case 1 : $\ell_t = r_t = i$

In this case the closed-loop state dynamics are

$$\dot{x}_t = (A_{ii} + B_{ii}\,\Gamma_i)\,x_t = \tilde{A}_{ii}\,x_t$$

and, noting that V does not depend explicitly on time, we have

$$\begin{aligned} E\,\{V(x_{t+\Delta}, \ell_{t+\Delta}, r_{t+\Delta}, t+\Delta) \mid x_t = x, \ell_t = i, r_t = i, t\} \\ = V(x, i, i, t) \\ + \Delta\, E\,\{\dot{x}_t' \mid x_t = x, \ell_t = i, r_t = i, t\}\; P_{ii}\,x \\ + \Delta\, x'\, P_{ii}\, E\,\{\dot{x}_t \mid x_t = x, \ell_t = i, r_t = i, t\} \\ + \Delta \sum_{j=1}^{N} \pi_{ij} V(x, i, j) + \Delta \sum_{j=1}^{N} \pi^1_{ij} V(x, j, i) + o(\Delta) \end{aligned}$$

and we obtain

$$\mathscr{L}V = x'\,(\tilde{A}_{ii}'\, P_{ii} + P_{ii}\tilde{A}_{ii} + \sum_{j=1}^{N} \pi_{ij}\, P_{ij} \;+ \sum_{j=1}^{N} \pi^1_{ij}\, P_{ji})\; x$$

or, using (7.4.3),

$$\mathscr{L}V = -\,x'Q_{ii}x \qquad \text{when } \ell_t = i,\ r_t = i \tag{7.4.4}$$

Case 2 : $\ell_t = j,\ r_t = i$

The closed-loop dynamics become

$$\dot{x}_t = (A_{ji} + B_{ji}\,\Gamma_j)\,x_t = \tilde{A}_{ji}\,x_t$$

and the average increment of V is

$$\begin{aligned} &E\,\{V(x_{t+\Delta}, \ell_{t+\Delta}, r_{t+\Delta}, t+\Delta) \mid x_t = x, \ell_t = j, r_t = i, t\} \\ &\quad = V(x, j, i, t) \\ &\quad + \Delta\,E\,\{\dot{x}_t' \mid x_t = x, \ell_t = j, r_t = i, t\}\;P_{ji}\,x \\ &\quad + \Delta\,x'\,P_{ji}\,E\,\{\dot{x}_t \mid x_t = x, \ell_t = j, r_t = i, t\} \\ &\quad + \Delta\,\pi_{ji}^0 V(x, i, i, t) - \Delta\pi_{ji}^0 V(x, j, i, t) + o(\Delta) \end{aligned}$$

so that we obtain

$$\mathscr{L}V = x'\,(\tilde{A}_{ji}'\,P_{ji} + P_{ji}\tilde{A}_{ji} + \pi_{ij}^0\,(P_{ii} - P_{ji}))\,x$$

or, using (7.4.3),

$$\mathscr{L}V = -\,x'Q_{ji}x \qquad \text{when } \ell_t = j,\ r_t = i \tag{7.4.5}$$

Therefore in both cases we have

$$\mathscr{L}V = -\,x'Q_{ji}x \qquad \text{when } \ell_t = j,\ r_t = i, \quad i, j = 1,N$$

and, for c_3 the smallest eigenvalue among the nN^2 (positive) eigenvalues of the Q_{ij}, i, j = 1, N we deduce

$$\mathscr{L}V \;\leq -\,c_3\,\|\,x\,\|^2$$

which is condition (iv) of theorem 2.3 and completes the proof. □

The coupled Liapunov equations (7.4.3) exhibit two distinctive features : couplings between the P_{ij}s are due to the jumps of the actual regime (π_{ij}) and to the jumps of the estimated regime (π^0_{ij} and π^1_{ij}) and the A_{ij}, $i \neq j$, matrices, entering through the equations for P_{ij}, $i \neq j$, describe the effect of linearization around a wrong nominal.

We note that a stability analysis ignoring the nonlinear nature of the underlying plant and the imperfections of the detection algorithm would result in N, instead of N^2, Liapunov equations

$$0 = \tilde{A}_i' P_i + P_i\tilde{A}_i + Q_i + \sum_{j=1}^{N} \pi_{ij} P_j$$

This can be recovered from (7.4.3) by letting $\pi^0_{ij} \to \infty$ and $\pi^1_{ij} \to 0$, i, j = 1, N (perfect detection) $P_{ij} = P_j$, i, j = 1, N, $A_{ii} = A_i$, i = 1, N and dropping the P_{ij}, $i \neq j$, equations.

<u>Example</u> :

To make the implications of theorem 7.2 more apparent, it is worth considering a simple scalar example with two modes

mode 1: $\dot{x} = f(x, u, 1)$

mode 2: $\dot{x} = f(x, u, 2)$

and symetric transitions $\Pi = \begin{pmatrix} -\pi & \pi \\ \pi & -\pi \end{pmatrix}$

For this scalar system the linearized dynamics are simply scalar coefficients a_{ij} and b_{ij}, i, j = 1, 2 and the detection and false alarm rates are taken as $\pi^0_{12} = \pi^0_{21} = \pi^0 = b\pi$ and $\pi^1_{12} = \pi^1_{21} = \pi^1 = a\pi$, for a and b positive numbers. With a perfect detection device one would get $b = \infty$ (instantaneous detection) and $a = 0$ (no false alarm), but, considering some of the examples treated in (Basseville and Benveniste, 1986), it was found that typical attainable values with classical algorithms are closer to $b \sim 10$ and $a \sim 5$. That is the detection is an order of magnitude faster than the actual jumps while there are on the average several (five in our example) false alarms between successive true transitions. Of course it would be possible to retune the detector sensitivity to obtain a lower false alarm rate but the delay would then increase and it will be seen next that only the ratio $(1 + a)/b$ affects the stability conditions.

Replacing upper-case letters by their lower-case counterparts, (7.4.3) becomes

$$0 = 2a_{11}\, p_{11} + 1 + \pi\, p_{12} - \pi\, p_{11} + \pi^1\, p_{21} - \pi^1\, p_{11}$$

$$0 = 2a_{22}\, p_{22} + 1 + \pi\, p_{21} - \pi\, p_{22} + \pi^1\, p_{12} - \pi^1\, p_{22}$$

$$0 = 2a_{12}\, p_{12} + 1 + \pi^0\, p_{22} - \pi^0\, p_{12}$$

$$0 = 2a_{21}\, p_{21} + 1 + \pi^0\, p_{11} - \pi^0\, p_{21}$$

for $q_{ij} = 1$, $i, j = 1, 2$.

To get a simple stability condition, the case where π becomes very large is now considered. When π is large the system alternates quickly between its two regimes but the relative time-scale of the detections and the false alarms is preserved through $\pi^0 = b\pi$ and $\pi^1 = a\pi$. The trick, which corresponds to a mixing of the hybrid dynamics by the regime Markov chain, was already employed in chapter 2 (see also (Geman, 1979)). In this limiting case, the above system yields a single stability condition for the closed-loop dynamics

$$\tilde{a}_{11} + \tilde{a}_{22} + (1/b)\,(1 + a)\,(\tilde{a}_{12} + \tilde{a}_{21}) \;<\; 0 \qquad (7.4.6)$$

From (7.4.6) several observations can be made:

- For the ideal case ($b = \infty$, $a = 0$), a condition similar to that of chapter 2 is recovered

$$\tilde{a}_{11} + \tilde{a}_{22} \;<\; 0$$

which shows that stability is ruled by "averaged dynamics": the system can be stable with one unstable regime (say $\tilde{a}_{11} > 0$) provided the other regime compensates ($\tilde{a}_{22} < -\tilde{a}_{11}$). This is again the phenomenon of chapter 2 and it is typical of the interactions between the continuous and discrete parts of hybrid systems.

- The obtained condition involves $\tilde{a}_{ij}$, $i \neq j$, and the detection and false alarm rates. Hence the condition is not satisfied even with $\tilde{a}_{11} + \tilde{a}_{22}$ negative if $\tilde{a}_{12} + \tilde{a}_{21}$ is positive enough (with a = 5 and b = 10 the threshold is $-(\tilde{a}_{11} + \tilde{a}_{22})$ less than $0.6(\tilde{a}_{12} + \tilde{a}_{21})$) so that this example demonstrates that *stabilization of the plant during quiescent periods* ($\tilde{a}_{11} + \tilde{a}_{22} < 0$) *is not sufficient to guarantee stability of the complete model.* Note that *it is not necessary either* since (7.4.4) can be satisfied with $\tilde{a}_{11} + \tilde{a}_{22}$ positive provided $\tilde{a}_{12} + \tilde{a}_{21}$ is stable enough ($-(\tilde{a}_{11} + \tilde{a}_{22}) > 0.6(\tilde{a}_{12} + \tilde{a}_{21})$)

- Imperfection of the detection device can produce instability in the adverse case where $\tilde{a}_{12} + \tilde{a}_{21}$ is positive and this can happen even when the detection is fast (b large) and the false alarms rare (a small). This shows that the detection delay and false alarm rate cannot be considered as the only figures of merit for an algorithm used in a reconfiguration loop and that attention must be paid to the influence of these imperfections on the stability of the reconfigured sysrtem. A design that does not account for these problems may stabilize either $\tilde{a}_{11} + \tilde{a}_{22}$ or $\tilde{a}_{11}$ and $\tilde{a}_{22}$ (see the two types of JLQ optimal stabilization discussed in chapter 3) but a new approach is desired to meet a condition like (7.4.4).

Of course this example is over simplified but it nevertheless captures the qualitative phenomena that can be expected in a more realistic situation and it clearly displays the role of the averaged dynamics, detection delays and false alarms.

7.5. Passively Adaptive Control Optimization

In § 7.2 the presence of small regime errors was analyzed and we proposed a regulator that guards itself against the uncertainty induced by clusters of hardly distinguishable regimes. We showed in § 7.4 that a controller ignoring false alarms and detection delays can fail to stabilize the system. Using the regime detection error model of §7.3 we propose here a regulator synthesis that explicitly, and quantitatively, accounts for the large regime errors.

As we have done previously, we shall pursue an optimal design, based on the minimization of a quadratic loss function. Contrary to § 7.2, we shall try to learn the current value of the regime to reduce uncertainty, but, to achieve an explicit solution, we

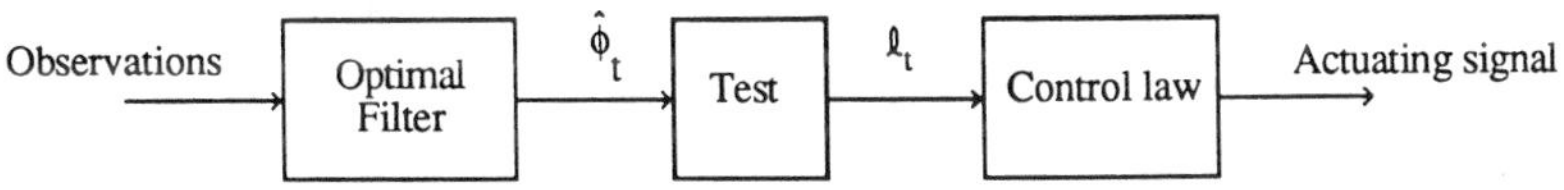

Figure 7.2 - Structure of the controller.

shall restrict attention to passive learning. The passivity restriction implies that the control signal has no direct impact on the rate at which the value of r_t is learnt. The more difficult active optimization problem is further discussed in § 7.6.

The filters developed in chapter 6 provide suitable regime estimators and we shall use them in cascade with regulators inhibited to probing actions. The general structure of the solution is then as follows

For the output of the detection test (ℓ_t) the model of § 7.3 is adopted. However the variable ℓ_t extracts only part of the information contained in $\hat{\phi}_t$. To be specific, suppose that a simple threshold test of level .9 is performed : it makes a difference to decide $\ell_t = 1$ because $\hat{\phi}_1 = .91$ or because $\hat{\phi}_1 = .999$ and decision is made with greater confidence in the second case. To capture this notion of confidence (or uncertainty) on ℓ_t we define a new variable $\tilde{\phi}_t$ as

$$\tilde{\phi}_t = \phi(\ell_t) - \hat{\phi}_t$$

where $\phi(\ell_t)$ is the indicator of ℓ_t. This vector can be computed at the output of the filter and test block and thus used in the control law as shown on figure 7.3

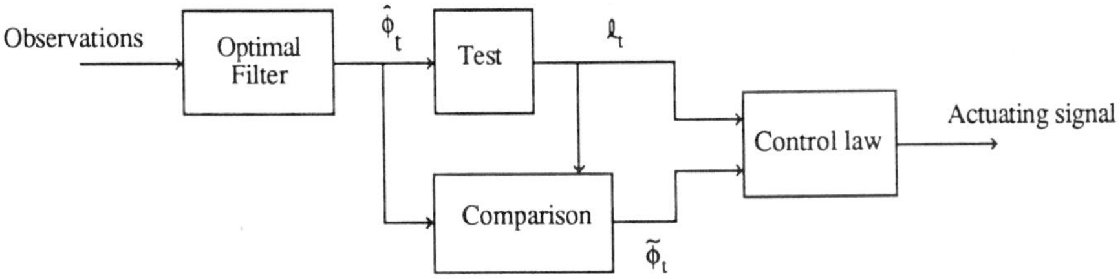

Figure 7.3 - Utilization of the confidence measure.

The measure of confidence (7.5.1) is clearly distinct from the estimation error $\phi(r_t) - \hat{\phi}_t$ which cannot be computed from the observations. The component of the vector (7.5.1) are small most of the time :

$$\begin{aligned} &\tilde{\phi}_{ti} = 1 - \hat{\phi}_{ti} \quad \text{and} \quad \hat{\phi}_{ti} \approx 1 \\ & \qquad\qquad\qquad\qquad\qquad \text{when } \ell_t = i \\ &\tilde{\phi}_{tj\neq i} = - \hat{\phi}_{tj} \quad \text{and} \quad \hat{\phi}_{tj} \approx 0 \end{aligned} \tag{7.5.2}$$

However, during a transient, an increase of $\tilde{\phi}_t$ provides an early warning of the jump, before $\hat{\phi}_t$ has crossed the detection threshold. Hence the idea of using $\tilde{\phi}_t$ to compensate the control law when ℓ_t is not really trusted (see figure 7.3).

The non-linear state dynamics being difficult to analyze some approximations are introduced as before. The deviations from the nominal set-point are defined

$$\begin{aligned} x_t &= x_{pt} - x_n(i) \\ u_t &= u_{pt} - u_n(i) \end{aligned} \tag{7.5.3}$$

Following a transition of r_t, say from i to j, the regulator tries to achieve operation close to $x_n(i)$ until ℓ_t switches to j. In other words, (7.5.3) should be read more precisely as

$$x_t = x_{pt} - x_n(\ell_t)$$

$$u_t = u_{pt} - u_n(\ell_t)$$

We distinguish between quiescent situations when the detection algorithm has correctly identified the mode ($\ell_t = r_t$, case 1) and transient ones ($\ell_t \neq r_t$, case 2) when there is a discrepancy between the perceived regime and the true one.

Consider first a quiescent situation.

Case 1 : $\ell_t = r_t = i$

This is the classical case and we define

$$A_{ii} = f_{x_p}\,(x_n(i),\, u_n(i),\, i)$$

$$B_{ii} = f_{u_p}\,(x_n(i),\, u_n(i),\, i)$$

The exact non linear model is approximated for small deviations by the linear equation

$$\dot{x}_t = A_{ii}\, x_t + B_{ii}\, u_t \tag{7.5.4}$$

Note however that jumps already cause a complication because, if the plant state x_{pt} is physically continuous, the state x_t of the linearized model (7.5.4) experiences discontinuities at instants when ℓ_t jumps. If $\ell_t = j \neq i = \ell_{t^-}$ then

$$\begin{aligned} x_t &= x_{pt} - x_n(\ell_t) \\ &= x_t + \delta\,(\ell_t, \ell_{t^-}) \end{aligned} \tag{7.5.5}$$

$$\text{with } \delta\,(\ell_t, \ell_{t^-}) = x_n(\ell_{t^-}) - x_n(\ell_t)$$

Consider next a transient situation.

<u>Case 2</u> : $\ell_t = i,\ r_t = j$

In this situation the regulator is not aware of the actual desired level $x_n(j)$ and it strives to maintain operation near what it perceives to be the desired level, $x_n(i)$. Defining

$$\begin{aligned} A_{ij} &= f_{x_p}\,(x_n(i),\, u_n(i),\, j) \\ B_{ij} &= f_{u_p}\,(x_n(i),\, u_n(i),\, j) \\ \Delta(i, j) &= f\,(x_n(i),\, u_n(i),\, j) \end{aligned} \tag{7.5.6}$$

the approximate linear model now is

$$\dot{x}_t = A_{ij}\, x_t + B_{ij}\, u_t + \Delta(i, j) \tag{7.5.7}$$

(note that the convention $\Delta(i, i) = 0$, $i = 1$ to N, makes it possible to write (7.5.4) as a special case of (7.5.7)).

In § 7.4 we used (7.5.7) omitting the Δ term and we neglected the δ discontinuity.

If the value of r_t could be determined without ambiguity, we would pick a controller like

$$u_{pt} = u_n(r_t) + \Gamma(r_t)\,(x_{pt} - x_n(r_t))$$

with the feedforward term cancelling the steady state influence of regime changes and the feedback term producing a corrective action to attenuate model errors and unmodelled disturbances. An optimal choice of Γ with respect to an average quadratic cost function was the subject of chapter 3. In this chapter the hypothesis of an exact measure of r_t is removed and the reconfiguration is based on the imperfect estimate ℓ_t.

The two sets of linearized models (7.5.4), (7.5.7) with the discontinuities (7.5.5) form our model for the state deviation dynamics. For ℓ_t we use the model of § 7.3 where $\hat{\phi}_t$ is generated by the one of filters of chapter 6. The equations of the filter depend on the available observations and we shall assume that there exists a set of good quality sensors providing a direct, albeit noisy, information on the current regime (typically the direct regime sensor (6.2.1)). Denoting by Y_t the corresponding σ-algebra , the complete information available at the controller is the product $X_t \vee Y_t$ where $X_t = \sigma - \{x_{ps}, t_0 \leq s \leq t\}$, in other words we also make the hypothesis that the state is measured.

Using $X_t \vee Y_t$ to estimate r_t leads to the dual control difficulty and the passive learning restriction is therefore mathematically enforced by making the approximation

$$\hat{\phi}_t = E\,\{\phi_t \mid X_t \vee Y_t\} \approx E\,\{\phi_t \mid Y_t\}$$

This will be valid if the direct sensors (Y_t) are good enough so that using measures of the state to estimate the regime would provide only marginal improvements.

To increase the flexibility of the design, the cost function considered so far is slightly generalized to include jump costs

$$J = E\{ \int_{t_0}^{t_f} g_c (x_{pt}, u_{pt}, \hat{\phi}_t, r_t)\, dt \tag{7.5.8}$$

$$+ \sum_{\substack{t_s \\ t_0 \le t_s < t_f}} \sum_{j=1}^{N} g_d (x_{pt_s}, u_{pt_s}, \hat{\phi}_{t_s}, r_{t_s}, j)\, \Delta z (r_{t_s}, j, t_s) \mid x_0, i_0, t_0\}$$

where z(i, j, t) is a counting process recording the number of jumps from i to $j \neq i$ that have occurred up to time t. In other words $\Delta z(r_t, j, t)$ is equal to zero except at the instants t_s ($t_0 \le t_s < t_f$) when the plant regime jumps.

The integral of the instantaneous cost g_c depends upon the mode r_t to allow a changing balance between state and control penalization: in a failure regime it may be desirable to spend large amounts of control to preserve safe operation while economic reasons may advise to weight more heavily control expenditures in the normal regime. The transition cost g_d is incurred when the regime actually jumps and the designer can use it to penalize state and/or control excursions during transient operation. In a fault-tolerance application it can for example be associated with a penalty for turning-on a stand-by component or with the expenses of calling a maintenance team.

The class of admissible control laws $\mathscr{U}$ is defined as $X_t \vee Y_t$-measurable functions satisfying our usual smoothness conditions (see appendix 2). The minimization of (7.5.8) over $u \in \mathscr{U}$ for a system described by (7.5.4), (7.5.5), (7.5.7) is a markovian decision problem and the formalism of dynamic programming can conveniently be used. The minimum cost-to-go functional is introduced as

$$V(t, x_{pt}, \hat{\phi}_t) = \min_{u_p} E\{ \int_{t}^{t_f} g_c (x_{pt}, u_{pt}, \hat{\phi}_t, r_t)\, dt$$

$$+ \sum_{\substack{t_s \\ t \le t_s < t_f}} \sum_{j=1}^{N} g_d (x_{pt_s}, u_{pt_s}, \hat{\phi}_{t_s}, r_{t_s}, j)\, \Delta z (r_{t_s}, j, t_s) \mid X_t \vee Y_t\}$$

$$V(t_f, -, -) = 0$$

We define

$$V(t, x_{pt}, \hat{\phi}_t, r_t) = \underset{u_p}{\text{Min}}\ E\ \{ \int_t^{t_f} g_c\ (x_{pt}, u_{pt}, \hat{\phi}_t, r_t)\ dt \tag{7.5.9}$$

$$+ \sum_{\substack{t_s \\ t \le t_s < t_f}} \sum_{j=1}^{N} g_d\ (x_{pt_s}, u_{pt_s}, \hat{\phi}_{t_s}, r_{t_s}, j)\ \Delta z\ (r_{t_s}, j, t_s) \mid X_t \vee Y_t, r_t\}$$

and, with the approximation $\hat{\phi}_t = E\ \{\phi_t \mid Y_t\}$,an equivalent expression for the cost to go is

$$V(t, x_{pt}, \hat{\phi}_t) = \sum_{i=1}^{N} V(t, x_{pt}\ \hat{\phi}_t, i)\ \hat{\phi}_{ti} \tag{7.5.10}$$

Note that we shall use the same notation (V) for various parameterizations of the cost-to-go.

Expending (7.5.9) at instant t for $r_t = i$ we get

$$V(t, x_{pt}\ \hat{\phi}_t, i) = g_c\ (x_{pt}, u_{pt}, \hat{\phi}_t, i)\ dt$$

$$+ E\ \{ \sum_{\substack{t_s \\ t < t_s \le t+dt}} \sum_{j=1}^{N} g_d\ (x_{pt_s}, u_{pt_s}, \hat{\phi}_{t_s}, r_{t_s}, i, j)\ \Delta z\ (i, j, t_s) \mid X_t \vee Y_t, r_t = i\}$$

$$+ E\ \{V(t + dt, x_{pt+dt}, \hat{\phi}_{t+dt}, r_{t+dt} \mid X_t \vee Y_t, r_t = i\}$$

But for dt small, the second term in the RHS contributes only one jump over :

$$\sum_{\substack{j=1 \\ j \ne i}}^{N} \pi_{ij}\ g_d\ (x_{pt}, u_{pt}, \hat{\phi}_t, i, j)\ dt + o(dt) \tag{7.5.11}$$

For the last expectation in the RHS the only term involving u_{pt} is

$$V'_{xp}\ (t, x_{pt}, u_{pt_s}, \hat{\phi}_t, i)\ dx_{pt} = \tag{7.5.12}$$

$$V'_{xp}\ (t, x_{pt}, u_{pt_s}, \hat{\phi}_t, i)\ f(x_{pt}, u_{pt}, i)dt$$

Using (7.5.12), (7.5.11) and (7.5.10) in the minimum cost-to-go functional it is seen that the optimal control u^*_{pt} is a solution of

$$\partial/\partial u_p \, [\sum_{i=1}^{N} \hat{\phi}_{ti} \, [g_c \, (x_{pt}, u_{pt}, \hat{\phi}_t, i) + \sum_{\substack{j=1 \\ j\neq i}}^{N} \pi_{ij} \, g_d \, (x_{pt}, u_{pt}, \hat{\phi}_t, i, j)$$
$$+ V_{xp}' \, (t, x_{pt}, u_{pt_s}, \hat{\phi}_t, i) \, f(x_{pt}, u_{pt}, i)]] = 0$$

The linearized approximations of f make an explicit solution possible, but it is also needed to restrict the costs g_c and g_d to be quadratic with

$$g_c(x_{pt}, u_{pt}, r_t) = \tag{7.5.13}$$
$$(x_{pt} - x_n(r_t))' \, Q_1(r_t) \, (x_{pt} - x_n(r_t)) + (u_{pt} - u_n(\ell_t))' \, Q_2(r_t) \, (u_{pt} - u_n(\ell_t))$$

and

$$g_d(x_{pt}, u_{pt}, r_t, j) = \tag{7.5.14}$$
$$(x_{pt} - x_n(r_t))' \, Q_{1j}(r_t) \, (x_{pt} - x_n(r_t)) + (u_{pt} - u_n(\ell_t))' \, Q_{2j}(r_t) \, (u_{pt} - u_n(\ell_t))$$

Note that the dependence of (7.5.8) on $\hat{\phi}_t$ has been exploited : $x_n(r_t)$ is used in (7.5.13), (7.5.14) because it is the true desired level, while $u_n(\ell_t)$ is used because r_t is not known and the control volatility is restricted with respect to the perceived nominal level.

For future use the following matrices are introduced

$$\bar{Q}_{1i} = Q_{1i} + \sum_{\substack{j=1 \\ j\neq i}}^{N} \pi_{ij} \, Q_{ij}(i)$$
$$\bar{Q}_{2i} = Q_{2i} + \sum_{\substack{j=1 \\ j\neq i}}^{N} \pi_{ij} \, Q_{2j}(i) \qquad , i = 1 \text{ to } N \tag{7.5.15}$$
$$\bar{\bar{Q}}_{2i}^{-1} = (I + \bar{Q}_{2i}^{-1} \sum_{\substack{j=1 \\ j\neq i}}^{N} \tilde{\phi}_{tj} \, Q_{2j}) \, \bar{Q}_{2i}^{-1}$$

With a high quality direct regime sensor a first order approximation in terms of the small $\tilde{\phi}_t$ parameter is retained to obtain an explicit near optimal solution

Theorem 7.3

Up to the first order in $\tilde{\phi}_t$, the optimal passive regulator for (7.5.4), (7.5.7), (7.5.5) and (7.5.8) is given by

$$u_t^* = -\bar{\bar{Q}}_{2i}^{-1} B_{ii}' (\Lambda_{ii} x_t + \lambda_{ii}) \tag{7.5.16}$$

$$+ \bar{Q}_{2i}^{-1} [\sum_{j=1}^{N} \tilde{\phi}_{tj} B_{ij}' (\Lambda_{ij} x_t + \lambda_{ij})]$$

where

$$-\dot{\Lambda}_{ii} = A_{ii}'\Lambda_{ii} + \Lambda_{ii} A_{ii} - \Lambda_{ii} B_{ii}'\bar{Q}_{2i}^{-1} B_{ii}\Lambda_{ii} + \bar{Q}_{1i} + \sum_{j=1}^{N} \pi_{ij} \Lambda_{ij} + \sum_{j=1}^{N} \pi_{ij}^{1} \Lambda_{ji} \tag{7.5.17}$$

$$-\dot{\Lambda}_{ij} = (A_{ij} - B_{ij} \bar{Q}_{2i}^{-1} B_{ii}'\Lambda_{ii})' \Lambda_{ij} + \Lambda_{ij} (A_{ij} - B_{ij} \bar{Q}_{2i}^{-1} B_{ii}'\Lambda_{ii}) + Q_i$$

$$+ \Lambda_{ii} B_{ii}'\bar{Q}_{2i}^{-1} Q_{2i} \bar{Q}_{2i}^{-1} B_{ii}'\Lambda_{ii} + \pi_{ij}^{0} (\Lambda_{jj} - \Lambda_{ij}) \tag{7.5.18}$$

$$-\dot{\lambda}_{ii} = (A_{ii} - B_{ii} \bar{Q}_{2i}^{-1} B_{ii}'\Lambda_{ii})' \lambda_{ii} + \sum_{j=1}^{N} \pi_{ij} \lambda_{ij} + \sum_{j=1}^{N} \pi_{ij}^{1} (\lambda_{ji} + \Lambda_{ji} \delta(j, i)) \tag{7.5.19}$$

$$-\dot{\lambda}_{ij} = (A_{ij} - B_{ij} \bar{Q}_{2i}^{-1} B_{ii}'\Lambda_{ii})' \lambda_{ii} + Q_i \delta(j, i) + \Lambda_{ii} B_{ii}' \bar{Q}_{2i}^{-1} Q_{2i} \bar{Q}_{2i}^{-1} B_{ii}'\lambda_{ii}$$

$$+ \pi_{ij}^{0} (\lambda_{jj} - \lambda_{ij}) - \Lambda_{ij} B_{ij} \bar{Q}_{2i}^{-1} B_{ii}'\lambda_{ii} + \pi_{ij}^{0} \Lambda_{jj} \delta(j, i)$$

$$+ \Lambda_{ij} \Delta(i, j) \tag{7.5.20}$$

with $\Lambda_{ii}(t_f) = \Lambda_{ij}(t_f) = 0$, $\lambda_{ii}(t_f) = \lambda_{ij}(t_f) = 0$, i, j = 1 to N.

The leading term of the corresponding cost is then

$$J = x_{t0}' (\Lambda_{ii}(t_0) - \sum_{j=1}^{N} \tilde{\phi}_{t0j} \Lambda_{ij}(t_0)) x_{t0} + 2x_{t0}' (\lambda_{ij}(t_0) - \sum_{j=1}^{N} \tilde{\phi}_{t0j} \lambda_{ii}(t_0))$$

$$+ \mu_{ii}(t_0) - \sum_{j=1}^{N} \tilde{\phi}_{t0j} \mu_{ij}(t_0)$$

and we omit the μ_{ij}, i, j = 1, N, equations.

Proof :

For quadratic costs, the optimality condition $\frac{\partial}{\partial u_p}(\cdot) = 0$ can be computed to give

$$0 = \sum_{i=1}^{N} \hat{\phi}_{ti} \left[2\bar{Q}_{2i} u_t^* + V'_{xp}(t, x_{pt}, \hat{\phi}_t, i) f_{u_p}(x_{pt} u_{pt}^*, i)\right] \qquad (7.5.21)$$

With the approximate linear models, f_{u_p} is independent of u_{pt}^* so that (7.5.21) is readily solvable. When $\ell_t = i$, the solution is

$$u_t^* = -0.5 \left(\bar{Q}_{2i} - \sum_{j=1}^{N} \tilde{\phi}_{tj}\bar{Q}_{2j}\right)^{-1} \left[B'_{ii} V_{x_p}(t, x_{pt}, \hat{\phi}_t, i) - \sum_{j=1}^{N} \tilde{\phi}_{tj} B'_{ij} V_{x_p}(t, x_{pt}, \hat{\phi}_t, j)\right] \qquad (7.5.22)$$

We remark that $x_{pt} = x_t + x_n(\ell_t)$ and that the pair $(\tilde{\phi}_t, \ell_t)$ contains the same information as $\hat{\phi}_t$, and a convenient reparameterization is $V = V(t, x_t, \tilde{\phi}_t, \ell_t, r_t)$. Letting V assume the following quadratic form

$$V(t, x_{pt}, \hat{\phi}_t, i, j) = x_t' \Lambda_{ij} x_t + 2x_t' \lambda_{ij} + \mu_{ij}$$

equation (7.5.22) becomes

$$u_t^* = -\left(\bar{Q}_{2i} - \sum_{j=1}^{N} \tilde{\phi}_{tj}\bar{Q}_{2j}\right)^{-1} \left[B'_{ii}(\Lambda_{ii} x_t + \lambda_{ii}) - \sum_{j=1}^{N} \tilde{\phi}_{tj} B'_{ij}(\Lambda_{ij} x_t + \lambda_{ij})\right]$$

when $\ell_t = i$

Up to first order terms $\tilde{\phi}_t$, this is simplified as

$$u_t^* = -\bar{\bar{Q}}_{2i}^{-1} B'_{ii}(\Lambda_{ii} x_t + \lambda_{ii}) + \bar{Q}_{2i}^{-1} \sum_{j=1}^{N} \tilde{\phi}_{tj} B'_{ij}(\Lambda_{ij} x_t + \lambda_{ij}) \qquad (7.5.23)$$

where the Q_2 matrices were defined in (7.5.15). What remains to be done is to bring back (7.5.23) into the minimal cost-go-to functional to obtain $3N^2$ equations for the matrices

Λ_{ij}, vectors λ_{ij} and scalars μ_{ij}, i, j = 1 to N. We separate again the two previous cases, namely quiescent and transient situations.

Consider first quiescent situation.

Case 1 : $\ell_t = r_t = i$

In this case

$$V(t, x_{pt}, \hat{\phi}_t, i, i) = (x_t' \bar{Q}_{1i} x_t + u_t' \bar{Q}_{2i} u_t)\, dt$$
$$+ E\{V(t+dt, x_{t+dt}, \tilde{\phi}_{t+dt}, \ell_{t+dt}, r_{t+dt}) \mid X_t \vee Y_t, r_t = i\}$$

under $u = u^*$.

In the quiescent situation, true regime jumps and false alarms may occur and the linearized dynamics are used with $x_t = x_{pt} - x_n(\ell_t) = x_{pt} - x_n(r_t)$, so that for the last term in the RHS we have

$$\begin{aligned} E\{V(t+dt, x_{t+dt}, \tilde{\phi}_{t+dt}, \ell_{t+dt}, r_{t+dt}) \mid X_t \vee Y_t, r_t = i\} &= \\ E\{V(t, x_t, \tilde{\phi}_t, \ell_t, r_t) \mid X_t \vee Y_t, r_t = i\} & \\ + E\{V_t\, dt + V_x'\, dx_t + V_{\tilde{\phi}}'\, d\tilde{\phi}_t \mid X_t \vee Y_t, r_t = i\} & \\ + E\{V(t, x_t, \tilde{\phi}_t, \ell_t, r_{t+dt}) \mid X_t \vee Y_t, r_t = i\} & \\ + E\{V(t, x_t + \delta(\ell_{t+dt}, \ell_t), \tilde{\phi}_t, \ell_{t+dt}, r_t) \mid X_t \vee Y_t, r_t = i\} & \end{aligned} \tag{7.5.24}$$

Note that when a false alarm occurs (the ℓ_{t+dt} term in (7.5.24)) V has to be evaluated at $x_t + \delta(\ell_{t+dt}, \ell_t)$ because $x_n(\ell_{t+dt}) \neq x_n(r_t)$. Using (7.5.24) into V under $u = u^*$ the final form of Bellman's equation is

$$\begin{aligned} 0 = (x_t'\bar{Q}_{1i} x_t + u_t' \bar{Q}_{2i} u_t + V_t + V_x' (A_{ii} x_t + B_{ii} u_t))\, dt & \\ + \sum_{j=1}^{N} \pi_{ij} V(t, x_t, \tilde{\phi}_t, i, j) + \sum_{j=1}^{N} \pi_{ij}^1 V(t, x_t + \delta(j, i), \tilde{\phi}_t, j, i))\, dt & \\ + E\{V_{\tilde{\phi}}'\, d\tilde{\phi}_t \mid X_t \vee Y_t, r_t = i\} & \end{aligned} \tag{7.5.25}$$

under $u = u^*$.

The $V_{\tilde{\phi}}'$ term in (7.5.25) has a contribution of order 2 and it is neglected. Using (7.5.23) for the optimal control and omitting all second order terms , like powers of x_t are finally collected to obtain

Quadratic term

$$0 = \dot{\Lambda}_{ii} + A_{ii}'\Lambda_{ii} + \Lambda_{ii} A_{ii} - \Lambda_{ii} B_{ii}' \bar{Q}_{2i}^{-1} B_{ii}\Lambda_{ii} + \bar{Q}_{1i} + \sum_{j=1}^{N} \pi_{ij} \Lambda_{ij} + \sum_{j=1}^{N} \pi_{ij}^{1} \Lambda_{ji}$$

Linear term

$$0 = \dot{\lambda}_{ii} + (A_{ii} - B_{ii} \bar{Q}_{2i}^{-1} B_{ii}\Lambda_{ii})' \lambda_{ii} + \sum_{j=1}^{N} \pi_{ij} \lambda_{ij} + \sum_{j=1}^{N} \pi_{ij}^{1} (\lambda_{ji} + \Lambda_{ji} \delta(j, i))$$

Constant term

$$0 = \dot{\mu}_{ii} + \sum_{j=1}^{N} \pi_{ij} \mu_{ij} \sum_{j=1}^{N} \pi_{ij}^{1} \mu_{ji} + \lambda_{ii}' B_{ii}' \bar{Q}_{2i}^{-1} B_{ii} \lambda_{ii} + \sum_{j=1}^{N} \pi_{ij}^{1} \delta(j, i)' \Lambda_{ji} \delta(j, i)$$

with $\Lambda_{ii}(t_f) = 0$, $\lambda_{ii}(t_f) = 0$ and $\mu_{ii}(t_f) = 0$, which gives (7.5.17) and (7.5.19) of the theorem and the equations for the μ_{ii}, i = 1, N that were omitted.

Consider next a transient situation.

<u>Case 2</u> : $\ell_t = i$, $r_t = j$.

This situation corresponds to the detection delay or to the recovery from a false alarm, and it was assumed that the plant regime r_t does not jump during this short time lapse. The only possible jump is thus a jump of ℓ_t from i to j, corresponding to the end of the transient, and this jump occurs on the average after $1/\pi_{ij}^{0}$ seconds. Only the differences between the two cases are emphasized. Equation (7.5.25) becomes

$$\begin{aligned} 0 = &[(x_t + \delta(j, i))' Q_{1i} (x_t + \delta(j, i)) + u_t' Q_{2i} u_t + V_t \\ &+ V_x' (A_{ij}x_t + B_{ij}u_t + \Delta(i, j)] dt \\ &+ \pi_{ij}^{0} [V(t, x_t + \delta(j, i), \tilde{\phi}_t, j, j) - V(t, x_t, \tilde{\phi}_t, i, j)] dt \end{aligned}$$

where we have used $x_{pt} - x_n(r_t) = x_{pt} - x_n(\ell_t) + \delta(j, i)$. Note the presence of the $\Delta(i, j)$ term which also accounts for the error made by the detection filter.

Again using (7.5.23) for u^* and proceeding as in the first case, equations (7.5.18) and (7.5.20) of the theorem are obtained from the quadratic and linear terms. The constant term gives a relation for μ_{ij}, $i \neq j$, and the last step is to collect the above expression to write the minimal cost

$$J = V(t_0, x_{pt0}, \hat{\phi}_{t0})$$

Denoting by i the initial regime estimate ($\ell_{t0} = i$), we get the expression given in the theorem which completes the proof. □

The complexity of the obtained solution is more apparent than real. The N coupled Riccati-like equation (7.5.17) are solved off-line together with their N^2 - N Liapunov-like counterparts (7.5.18). Similarly (7.5.19) and (7.5.20) are N^2 coupled vector differential equations depending only on model data. Computations can also be simplified for large t_f, replacing (7.5.17) to (7.5.20) by algebraic equations obtained when setting the time derivatives to zero. Note that the matrices and vectors resulting from the solution of (7.5.17) to (7.5.20) actually reflect the previous modeling effort in that they depend on the jump detection black box model (through Π^0 and Π^1) and on the linearization error characteristics (δ and Δ). Once the Λ_{ij}'s and λ_{ij}'s, i, j = 1 to N, are known, (7.5.16) is easily implemented on a real-time computer.

The first noticeable property of (7.5.16) is its mixed, feedback and feedforward, structure. This will not be a surprise for practitioners, but in the research community it seems that the feedback aspect of control has attracted most attention while the use of feedforward signals has been neglected. This is probably due to the ubiquitous LQG theory which followed the work of Kalman and its success in many applications. However when the system experiences large and unpredictable changes in its operating conditions, for example because of failures, it is no longer possible to presuppose a dynamically passive, foreknown, nominal input. Indeed a feedforward control action then has to be designed which is the primary signal causing the plant state to move from one equilibrium to the other.

It is interesting to remark that with the proposed approach the feedforward and feedback terms emerge together naturally from a single optimizing design. The feedback terms in (7.5.16) depend on the Λ matrices while the feedforward terms come with the bias vectors λ. More is said about this structure in the discussion of the example below.

The second main characteristic of (7.5.16) is that both the most plausible regime ℓ_t and the uncertainty vector $\tilde{\phi}_t$ are used to adapt the control action. Indeed once the detection filter decides that a jump has occurred, that is once ℓ_t is switched, say from i to k, all the coefficients are reconfigured to their value with k in place of i. During quiescent situations $\tilde{\phi}_t$ is very small so that (7.5.16) is approximately equal to

$$u_t^* = -\bar{Q}_{2i}^{-1} B_{ii}^{'} (\Lambda_{ii} x_t + \lambda_{ii}) \qquad \text{when } \ell_t = i \tag{7.5.26}$$

which will regulate x_{pt} near $x_n(i)$ as long as no transition occurs. Note that due to the λ_{ii} term, the steady state value of x_t is not zero, and there is a bias, which corresponds to an anticipation of the eventual transition to another equilibrium. When a jump occurs, $\hat{\phi}_t$ takes larger values and, before the transition can be detected, ℓ_t switched to a new value and (7.5.26) reconfigured, the second term in (7.5.16)

$$\bar{Q}_{2i}^{-1} \sum_{j=1}^{N} \tilde{\phi}_{tj} B_{ij}^{'} (\Lambda_{ij} x_t + \lambda_{ij}) \qquad \text{when } \ell_t = i$$

provides an early reaction with a continuous adaptation.

<u>Example</u> :

To illustrate the previous discussion a simple, academic, example is considered in some detail. It is a scalar system (n = 1) with two regimes (N = 2) and the state dynamics includes a non linear actuator and a non linear loop

$$\dot{x}_{pt} = a + bx_{pt} + cx_{pt}^2 + dx_{pt}^3 + (e - x_{pt})\, u_{pt} \tag{7.5.27}$$

where the a to e coefficients are regime dependent. Figure 7.4 shows a block diagram of the plant with the proposed control system.

It is desired to regulate x_{pt} near $x_n(i) = 1.$ when $r_t = 1$ and near $x_n(2) = 2.$ when $r_t = 2$.

The direct regime sensor (see § 6.2 and figure 7.3) delivers

$$dy_t = r_t dt + \sigma dw_t$$

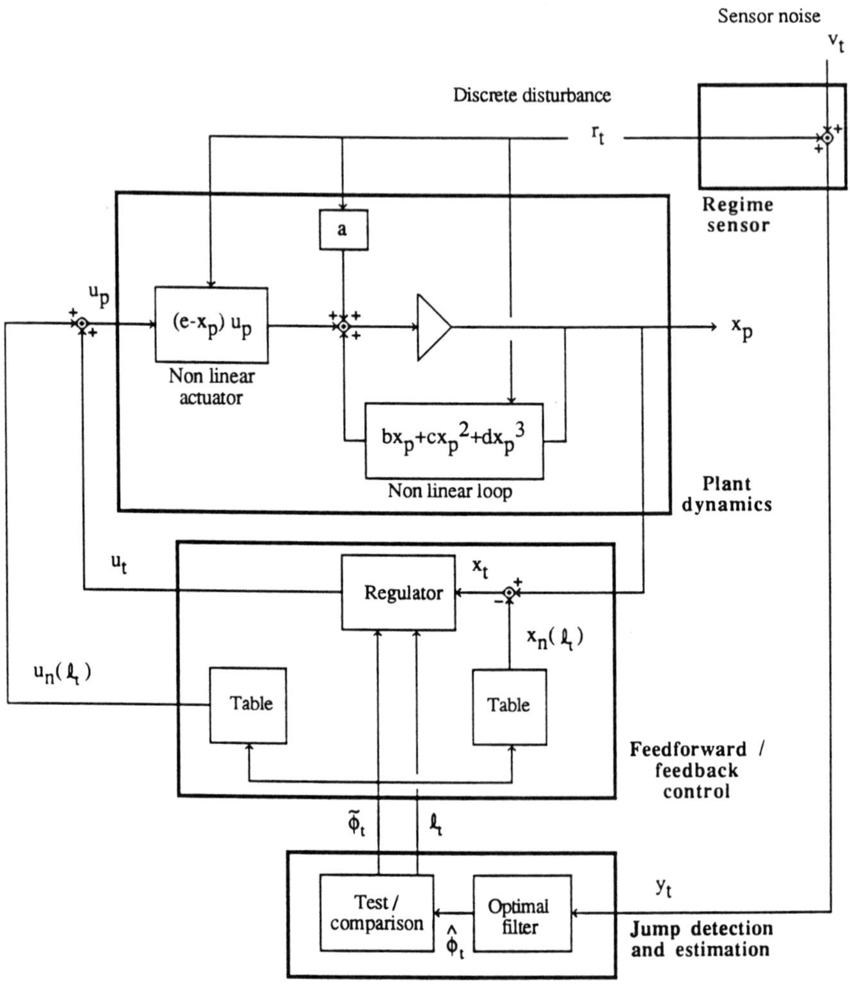

Figure 7.4 - Block diagram of the example.

with $\sigma = 0.1$ which makes it legitimate to base $\hat{\phi}_t$ solely on Y_t. The parameter values in (7.5.27) are chosen as

$$r_t = 1 \quad a_1 = -83. \quad b_1 = 190. \quad c_1 = -124. \quad d_1 = 27. \quad e_1 = 0.$$

$$r_t = 2 \quad a_2 = -17. \quad b_1 = -4. \quad c_2 = 11. \quad d_2 = 27. \quad e_2 = 3.$$

and the equilibrium values of u_{pt} are $u_n(1) = 10.$ and $u_n(2) = 5.$

The regime transition matrix is

$$\Pi = \begin{bmatrix} -0.25 & 0.25 \\ 0.25 & -0.25 \end{bmatrix}$$

The normalized version of the regime estimation filter $\hat{\phi}_t = E\{\phi_t \mid Y_t\}$ with the equation of theorem 6.1 produces the sample path of $\hat{\phi}_{t1}$ plotted on figure 7.5 when r_t is switched from 1 to 2 at $t = 4.$ and back to 1 at $t = 7.$

Matching the sample mean detection delay and mean time between false alarms, Π^0 and Π^1 were fixed as

$$\Pi^0 = \begin{bmatrix} -7. & 7. \\ 7. & -7. \end{bmatrix} \quad \Pi^1 = \begin{bmatrix} -0.75 & 0.75 \\ 0.75 & -0.75 \end{bmatrix}$$

It is interesting to compare the proposed solution to two natural competitors. The first one is the solution where the occurrence of false alarms is neglected. It is simply obtained from (7.5.17) to (7.5.20) by setting $\Pi^1 = 0$ (no false alarm). A second competitor is the solution which uses the same learning block for $\hat{\phi}_t$ but disregards the uncertainty information $\tilde{\phi}_t$. This controller is called myopic because it believes that ℓ_t is exactly identical to r_t without regard for short term misclassifications. Its equations are deduced as a special case of (7.5.16) to (7.5.20) when $\tilde{\phi}_t$ is set to zero (uncertainty is neglected), Π^1 is set to zero (no false alarm) and Π^0 to "∞" (infinitely fast detection). The result is

$$u_t(\text{myopic}) = -\bar{Q}_{2i}^{-1} B_{ii}' (\Lambda_{ii} x_t + \lambda_{ii}) \qquad \text{when } \ell_t = i$$

where

$$- \dot{\Lambda}_i = A_{ii}'\Lambda_i + \Lambda_i A_{ii} - \Lambda_i B_{ii} \bar{Q}_{2i}^{-1} B_{ii}'\Lambda_i + \bar{Q}_{1i} + \sum_{j=1}^{N} \pi_{ij} \Lambda_j$$

$$- \dot{\lambda}_i = (A_{ii} - B_{ii} \bar{Q}_{2i}^{-1} B_{ii}'\Lambda_i)' \lambda_i + \sum_{j=1}^{N} \pi_{ij} (\lambda_j + \Lambda_j \delta(j, i))$$

with $\Lambda_i(t_f) = 0$, $\lambda_i(t_f) = 0$, i = 1 to N.

With respect to the fault-tolerant control application, this myopic solution corresponds to the reconfiguration of the control law which is obtained if one neglects the imperfections of the failure detection device. In that sense it also corresponds to a certainty equivalence solution where the regime estimate is treated as if it were exact. On the contrary (7.5.16) to (7.5.20) enforce separation by splitting the estimation and regulation tasks between two devices, but, thanks to the model of § 7.3, the inevitable detection delays and false alarms are acknowledged.

The three regulators are compared on figure 7.6 for the sample path of $\hat{\phi}_t$ given on figure 7.5. Following the jumps of r_t, it is desired to track $x_n = 1$ up to t = 4., $x_n = 2$ from t = 4. to 7., then $x_n = 1$ again.

The first observation is the large improvement obtained over the myopic solution during the transients after t = 4. and t = 7. This is all the more remarkable as the detection is here pretty fast (see Π^1 compared to Π and figure 7.5): even though the detection delay is short and the false alarm rate low, the performance of the myopic control law is very poor. This in accord with the findings of § 7.4 where it was shown that myopic like reconfigurations can even destabilize the closed-loop dynamics.

Contrasting with the myopic solution, the two other solutions are able to respond quickly to changes in the operating conditions and avoid large deviations after t = 4. and t = 7. This is of course the benefit of the $\tilde{\phi}_t$ link from the estimation block to the control one on figure 7.4 that quickly informs them that something has happened. This link is severed in the myopic solution and it deprives it of the capacity to anticipate the ℓ_t decisions. Until t ~ 4.3 when ℓ_t is switched from 1 to 2, the myopic action tries to keep

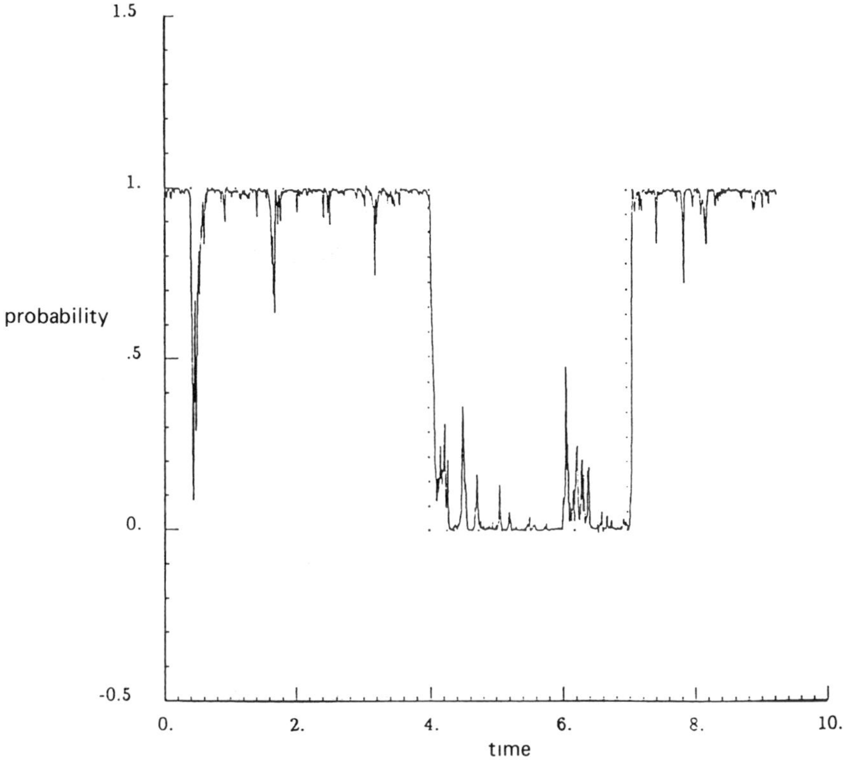

Figure 7.5 - Trajectory of the regime estimate.
(... ϕ_{t1}, —— $\hat{\phi}_{t1}$)

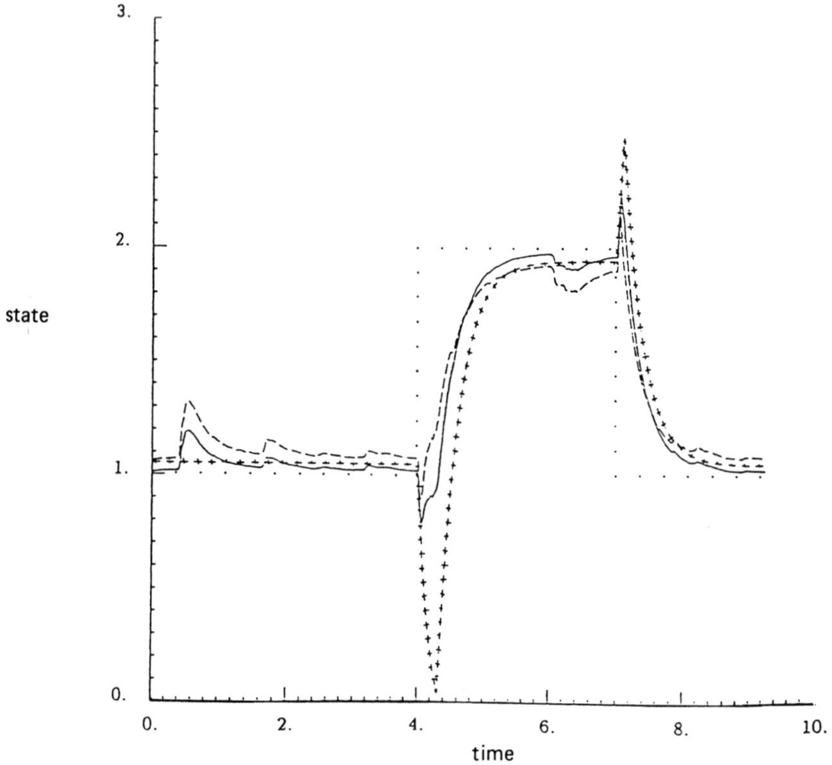

Figure 7.6 - State trajectories.
(+++ myopic solution, --- without false alarm modelling, —— proposed solution)

x_{pt} near $x_n(1)$ while the dynamics are in fact under $r_t = 2$. It simply happens that this ill conceived action is destabilizing. Once the correct decision is made, the myopic solution moves x_{pt} towards the right x_n but it is never quite able to make up its poor start before the next jump occurs.

A second observation concerns quiescent periods. At first look it seems that the myopic solution performs better in these situations. Indeed the $\tilde{\phi}_t$ multiplied signals in (7.5.16) make it sensitive to the noise in the regime sensor which is present, though filtered, in $\tilde{\phi}_t$. Hence while the myopic solution, ignoring $\hat{\phi}_t$, results in a flat trajectory from t = 0. to 4., the other trajectories are perturbed around t = 0.4 when figure 7.5 reveals that $\hat{\phi}_{t1}$ almost went to zero even though r_t remained equal to 1. This volatility during quiescent periods is somehow the price one has to pay to obtain a fast reaction to regime transitions. However a model warning the regulator against false alarms should be able to attenuate the phenomenon. And indeed the superiority of the proposed solution, where such a model is included via Π^1, is then understood: at the expense of a slightly slower response at t = 4. and t = 7., the volatility during the [0., 4.] interval is significantly reduced.

The third observation regards the bias. Remember that all three solutions generate biases in anticipation of the next transition (hence x_{pt} settles slightly above 1. = $x_n(1)$ over [0., 4.] and slightly below 2. = $x_n(2)$ over [4., 7.]). Of course an integral action would cancel these offsets but also slow down the transient dynamics. The interesting finding here is that the regulator of theorem 7.3 generates a smaller bias than the myopic one (look for example at the interval [2., 4.]). The reason is that it "knows", through the design model, that it is from time to time the victim of false alarms that drive x_{pt} further up towards $x_n(2) = 2.$ and, in anticipation of that, it lowers its steady state level closer to 1. So that in truth, during most of the quiescent periods, its deviation to the desired nominal is smaller than that of the myopic solution. It can therefore be claimed that the proposed regulator is superior to the myopic one even during quiescent intervals.

7.6 Notes and References

Control of hybrid systems is most difficult when the regime is not measured. Filters and detectors are used to estimate the missing variable but it is nevertheless necessary to design regulators that account for regime uncertainty.

In this chapter we have proposed solutions by distinguishing two types of uncertainty : a cluster of seemingly close regimes and large but transient errors of the decision logic. In the first case it is convenient to average over the cluster and the regulator is thus insensitive to less conspicuous regime differences. For false alarms or detection delays we introduced a model of the filter/detector performance that incorporates quantities like the average delay and the false alarm rate. We have demonstrated that a feedback law ignoring these errors can destabilize the system. A refined solution, utilizing both a reconfiguration and a continuous adaptation, is therefore advocated and we have illustrated its performance on an example.

Three findings of this chapter are also of interest on a more general level :

- When designing a regulator for a non linear system with abrupt parameter changes, the linearization procedure requires special care because of situations where there is a discrepancy between the perceived operating condition and the real one. Also the dependency of the equilibria on the jump parameter introduces discontinuities of the error state.

- The use of a high quality direct sensor for the regime permits a legitimate elimination of the dual control difficulty. When possible, this sensor should be asked for at the plant conception stage: the insolation sensor for the solar thermal receiver is an example but one may also think about built-in test procedures for the components of a fault-tolerance application.

- A model of rupture detection algorithms was proposed. It captures in a quantitative way the main phenomena, detection delays and false alarms, in a black box representation. It is important to stress that this is operative regardless of what precise scheme one chooses in the vast detection literature and, though quite crude, it is surely an improvement over the "perfect detection" alternative.

To arrive at tractable optimization problems we had to make a number of simplifying assumptions. For example the validity of the two sets of linearized state dynamics introduced in § 7.4 is not always guaranteed. It seems sound on an engineering basis but one could certainly find examples where less satisfactory performance would result.

However the most important limitation of our solutions is that they are constrained to avoid the dual control problem. It is interesting to discuss this difficulty further, referring to the situation of § 7.5. Allowing dual, or actively adaptive, control laws would amount to feeding x_{pt} in addition to y_t in the filter block of figure 7.4 and a dual solution would generate control signals near the jumps of r_t to improve the convergence of $\hat{\phi}_t$. Of course only two regimes were considered in the example and the detection was already pretty fast and there would not be much to gain in adding extra probing signals. But for a system with, say, three regimes such that the information delivered by the regime sensor poorly distinguishes between $r_t = 2$ and $r_t = 3$ (as encountered in § 6.2), we have seen that $\hat{\phi}_2$ and $\hat{\phi}_3$ remain both close to 0.5 following a jump of r from 1 to 2 (see figure 6.4). It might then be worth using a dual solution to excite the system and exploit, through X_t, a better separation of regimes 2 and 3 in the state dynamics. Indeed we have illustrated this in § 6.2 using linear state dynamics (compare figure 6.5 to figure 6.4). In the simulation of this example, fixed arbitrary excitations were added to the state deviation at every jump instant, but a dual regulator could do this in a feedback form. To turn this argument around one more time, it may however be noticed that following a jump of r_t the regime dependence of the equilibrium produces some excitation anyway. In such a situation the marginal improvement of a full dual solution has to be cautiously weighted against the accompanying increase in complexity. A useful tool is then the JLQ solution of chapter 3 with state and regime feedback. Assuming a complete knowledge of the plant variables (x_{pt} and r_t) the optimal cost can be computed and compared, for example, to the passive near optimal regulator of § 7.5. If the two costs are close (say with a suboptimality degree of 10 %) it will often be decided that the active solution is not worth the extra complexity.

There are applications however where it is known that probing can significantly improve performance, see for example the discussion of (Tse and Bar-Shalom, 1976) for a target tracking problem.

Given the difficulty of the exact dual optimization problem, a lot of efforts have been devoted to suboptimal solutions that preserve some of the properties of the optimal solution. We have already mentioned probing; knowing that future observations will be made an active controller anticipates benefits of improved estimation and regulates its adaptation for enhanced learning. In addition to its influence on the current state, active adaptation takes into account the future uncertainty. However excessive probing cannot be allowed because it is expensive in control resources and because there is a risk that excitation will increase the weight of uncontrollable uncertainties in the expected cost. This leads a cautious controller to moderate its action. Indeed the basic line of research in the dual control literature is to look for decompositions of the cost into caution and probing terms.

The dual control concept is traced back to (Feld'baum, 1965) and a survey on the problem is (Tse and Bar-Shalom, 1976). There have been many attempts on this subject including (Bar-Shalom and Tse, 1974, Saridis and Dao, 1972, Tse and Bar-Shalom, 1975) and research is still going on (e.g. Dersin, Athans and Kendrick, 1981, Sternby, 1978, Mookerjee, Bar-Shalom and Molusis, 1985). In the continuous time setting of the present book, the problem is best formulated along the lines of Rishel's work (1975, 1981) where the case of random parameter jumps is considered. Dynamic programming is preferred to arrive at suboptimal solutions by decomposition of the cost-to-go but the relationship with the maximum principle is also worth exploring (Rishel, 1978). A recent effort is (Caines and Chen, 1985) where a representation of the solution is obtained in terms of an auxiliary partial differential equation. This equation should be integrated on-line and, even if it were known how to solve it, it would severely tax the capacity of currently envisionned control computers. Also there is the formulation of (Hijab, 1987) where the cost is modified by an entropy additive term that quantifies the informational value of the solution. Directly related to hybrid models is the work of Loparo and his students, considering the JLQ problem with partial observations in discrete-time (e.g. Griffiths and Loparo, 1985, Casiello and Loparo, 1985). Decomposing the cost, minimization of a dual term in the decomposition results in probing signals that facilitate estimation at instants when uncertainty on the parameters is high.

The treatment of cluster of regimes that we proposed in § 7.2 evolved from (Sworder, 1984) while the rest of the chapter is based on original work (Mariton, 1987a and b, 1988) pursuing a research initiated in (Chou 1985, Sworder and Chou, 1985).

References

Bar-Shalom, Y., and Tse, E. (1974). Dual effect, certainty equivalence and separation in stochastic control, IEEE Trans. Aut. Control, AC-19 : 494.

Basseville, M., and Benveniste, A., Eds. (1986). Detection of Abrupt Changes in Signals and Dynamic Systems, Lecture Notes on Control and Information Sciences, Springer-Verlag, Berlin.

Blom, H.A.P. (1986). Overlook potential of systems with markovian coefficients, Proc. 25th IEEE Conf. Decision Control, Athens, pp. 1758-1764.

Caines, P.E., and Chen, H.F. (1985). Optimal adaptive LQG control for systems with finite state process parameters, IEEE Trans. Aut. Control, AC-30 : 185.

Casiello, F., and Loparo, K.A. (1985). A dual controller for linear systems with random jump parameters, Proc. 24th IEEE Conf. Decision Control, Fort-Lauderdale, pp. 911-915.

Chou, S.D. (1985). Passively adaptive regulator design for systems with multiple models. PhD Dissertation, Univ. California, San Diego.

Dersin, P.L., Athans, M., and Kendrick, D.A. (1981). Some properties of the dual adaptive stochastic control algorithm, IEEE Trans. Aut. Control, AC-26 : 1001.

Feld'baum, A.A. (1965). Optimal Control Systems, Academic Press, New-York.

Geman, S. (1979). Some averaging and stability results for random differential equations, SIAM J. Appl. Math., 36 : 86.

Griffiths, B.E., and Loparo, K.A. (1985). Optimal control of jump linear quadratic gaussian systems, Int. J. Control, 42 : 791.

Hijab, O.(1987). Stabilization of Control Systems, Springer-Verlag, New-York.

Kats, I.I., and Krasovskii, N.N. (1960). On the stability of systems with random attributes, J. Appl. Math. Mech., 24 : 1225.

Mariton, M.(1987a). Control of systems with imperfectly detected markovian model changes, Proc. 21st Asilomar Conf. Signals, Systems and Computers, Pacific Grove, pp.266-270.

Mariton, M. (1987b). Control of non linear systems with markovian parameter changes, submitted for publication.

Mariton M. (1988). Detection delays, false alarm rates and the reconfiguration of control systems, to appear Int. J. Control.

Mookerjee, P., Bar-Shalom, Y., and Molusis, J.A. (1985). Dual control and prevention of the turn-off phenomenon in a class of mimo systems, Proc. 24th IEEE Conf. Decision Control, Fort-Lauderdale, pp. 1888-1893.

Rishel, R. (1975). Control of systems with jump markov disturbances, IEEE Trans. Aut. Control, AC-20 : 241.

Rishel, R. (1978). The minimum principle, separation principle and dynamic programming for partially observed jump processes, IEEE Trans. Aut. Control, AC-23 : 1009.

Rishel, R. (1981) A comment in a dual control problem, IEEE Trans. Aut. Control, AC-26 : 606

Saridis, G.N., and Dao, T.K. (1972). A learning approach to the parameter adaptive self-organizing control problem, Automatica, 9 : 589.

Sternby, J.(1978). A regulator for time varying stochastic systems, Proc. 7th IFAC World Congress, Helsinki.

Sworder, D.D. (1984). Control of systems subject to small measurement disturbances, Trans.ASME, J.Dyn. Systems. Meas. Control, 106 : 182.

Sworder, D.D., and Chou, S.D. (1985). A survey of design methods for random parameter systems, Proc. 24th IEEE Conf. Decision Systems, Fort-Lauderdale, pp. 894-899.

Tse, E., and Bar-Shalom, Y. (1975). Generalized certainty equivalence and separation in stochastic control, IEEE Trans. Aut. Control, AC-20 : 817.

Tse, E., and Bar-Shalom, Y. (1976). Actively adaptive control for non linear stochastic systems, Proc. IEEE, 64 : 1172.

8

Extensions and Open Problems

8.1 Introduction

Throughout the previous chapters, the hybrid model introduced in chapter 1 was studied in depth and the body of results presented forms a basis for the solution of applied control and filtering problems where a discrete random variable interacts with a continuous one. Indeed, the refinements of the model that we suggested here and there were motivated by the needs of a particular situation. It is also true, however, that real behaviours sometimes exhibit qualitative features that were not well reflected so far and the purpose of this last chapter is to portray, succinctly, the various directions in which the model can, and should, be extended. We shall also point out some open theoretical problems that might attract researchers' attention.

Three extensions of the basic model are discussed in some detail:

- When we describe the passage of clouds over the heliostats in the solar thermal receiver the markovian assumption captures only a first-order approximation of the phenomenon. A closer look at meterological data suggests that a more refined model would include non markovian transitions to come closer to the typically gregarious pattern of cloud motion. The purpose of §8.2 is to present such a model and to illustrate its use by solving a filtering problem.

- The state dynamics considered so far have an intrinsic low-pass character with less zeros than poles, according to the limited bandwidth of physical systems. There are situations, however, where there exists a fast control-to-state influence and §8.3 is devoted to the corresponding wide band hybrid models. We discuss the effect of the larger system bandwidth on the behaviour of an optimal controller.

- The classical formulation of optimal control problems concerns a single decision-maker faced with a plant model, a control objective and design constraints. In §8.4 we consider the multiple decision-makers optimization problem. For example, in an economic planning application we imagine each economic agent as an individual decision-maker with a proper objective. Hierarchical relationships appear and disappear as stronger agents assume a leadership or decide to cooperate within a team. The hybrid aspect is interesting because it permits a description of sudden strategy changes as when one player loses or gains a leadership or when new teams are formed. The randomness of these transitions captures the uncertain and unpredictable environmental parameters.

8.2 Non Markovian Jumps

The markov chain model of regime transitions used so far corresponds to exponentially distributed sojourn times. There are situations where this is too restrictive: for a fault-tolerance application, basics from reliability theory suggest that the constant failure rate of the exponential distribution should be replaced by the "bath tub" curve, with higher rates near the end and beginning of a component life-time. This is due to aging and burn-in phenomena. For the tracking problem described in chapter 1 it is also useful to have some flexibility to fit the sojourn time distributions to observed behavioural patterns of the maneuvering target.

A direct approach is to explicitly write the dependence of the transition rates on the time elapsed since the last transition. Defining, for $r_t = i$, $\tau_t = \inf\{s \in [0,\infty[\mid r_{t-s} \neq i\}$, we consider Π as a function of τ_t. The transition probabilities are then

$$\mathscr{P}\{r_{t+\Delta} = j \mid r_t = i, \tau_t = \tau\} = \pi_{ij}(\tau)\Delta + o(\Delta) \quad \text{for } i \neq j \tag{8.2.1}$$

The regime is no longer a markov process. Processes of this type are often called semi-markov processes and are also related to doubly stochastic point processes. Assuming perfect observations, i.e. the controller knows the plant state x_p, the regime r_t and the length of time since the last jump τ_t, (Sworder, 1980) considered an optimal regulation problem for (8.2.1) with linear state dynamics and average quadratic cost. It is possible to write the Hamilton-Jacobi-Bellman equation and, using our usual notations, one is led to a cost-to-go like $x'\Lambda_i(\tau)x$. Unfortunately this produces a partial difference equation for the Λ_is with the partials $\partial\Lambda_i/\partial t$ and $\partial\Lambda_i/\partial\tau$. This PDE moreover exhibits peculiarities with variations at a point (t, τ) dependent on the value at $(t, 0)$. Standard numerical methods for

PDE are thus not readily applicable. Transforming the equation into an ordinary differential along a characteristic curve simplifies the expression but the anomaly of the original PDE reappears as unclassical boundary conditions. It is not easy to define reasonable approximations and this direct approach seems restricted to simple models.

These difficulties arise because of the generality of the model (8.2.1). In many instances we would be willing to give up some generality to arrive at a tractable design.

A simple way to describe non-markovian regime jumps is to consider the regime r_t as the projection, or the output, of a large markov chain.

Calling m_t the mode we assume that m_t is a markov chain on $S^o = \{1, 2, \ldots, N^o\}$ with generator Π^o and $N^o \geq N$. We partition S^o into N classes $Cl_1, Cl_2, \ldots, Cl_N$ and the regime is given by the projection

$$r_t = i \quad \text{when } m_t \in Cl_i, \quad i = 1, N. \tag{8.2.2}$$

To the markovian transitions of m_t, (8.2.2) associates non-markovian transitions for r_t and, choosing N^o and the generator $\Pi^o = (\pi^o_{ij})_{i,j\ =\ 1,\ N^o}$ for the mode, different jump distributions can be obtained as illustrated by the following example.

<u>Example</u>:

Single failure systems are considered, that is $S = \{1, 2\}$, $r_t = 1$ for the nominal regime and $r_t = 2$ for the failed one (the partition of S^o is then Cl_1 for the odd modes and Cl_2 for the even ones). First recall that for N = 2 with π_f as the failure rate, the markovian model leads to an exponential distribution for the lifetime of the nominal regime, with density $D_\tau = 1 - \exp(-\pi_f \tau)$.

Taking $N^o = 2L$ and Π^o band-diagonal with $\pi^o_{2k+1,2k+3} = -\pi^o_{2k+1,2k+3} = \pi^o_f$ for k = 0 to L - 2 and $\pi^o_{2L-1,2L} = -\pi^o_{2L-1,2L-1} = \pi^o_f$, the mode chain transits inside Cl_1 ($m_t = 1 \rightarrow 3 \rightarrow 5\ldots$) until it reaches $m_t = 2L-1$ where it jumps into Cl_2 ($m_t = 2L-1 \rightarrow 2L$) corresponding to the failure occurrence ($r_t = 1 \rightarrow 2$). The life time distribution of the nominal regime is then the gamma distribution, sum of L exponential distributions with rate π_f^o. This distribution was proposed by (Mortensen, 1987) to model the distribution of the on/off thermostat cycles in a heating system (see also Sworder and Elliott (1986) for a related renewal model of the background noise due to nuclear events).

This example shows that the mode would indeed be interpreted as an internal status of the system where degradation, even if it is not translated immediately as an external regime failure ($r_t = 1 \rightarrow 2$), prefigures a change in the system operating condition (one might think of the drift of a power source current as an anticipated signal of the component failure). Mode estimation would then be helpful for maintenance purposes: when a jump of the mode corresponding to an increased vibration level or a higher power consumption is detected, the operator may decide a preventive inspection of the system or even some anticipative maintenance.

Selecting other structures for the mode Markov chain, other distributions could have been obtained and indeed this model is very rich, as shown by the following approximation lemma, adapted from (Rudemo, 1973).

Lemma:

For any distribution of the nominal regime life-time, there exists a Markov chain m_t on $\mathbf{S}^o$ with a single absorbing mode such that this distribution is approximated arbitrarily close in Levy metric by the distribution of the time to absorption of m_t.

Proof:

Refer to Rudemo's paper (p. 602) with the Levy distance between distributions given in (Feller, 1966, p. 277). The idea is to actually construct the approximating chain, first decomposing the distribution into a weighted sum of step functions and then approaching each of the step functions by a sum of exponentials. □

As intuition suggests, and as Rudemo's construction demonstrates, it will be necessary to work with rather large mode Markov chains (N^o large) to describe adequately distributions notably different from the exponential. Some specific techniques might then be needed to handle this dimensionality problem.

To adapt the control law to mode jumps, the information contained in the measurements is exploited. We assume that the regime r_t is observed and denote as before

ϕ_t the mode indicator ($\phi_{ti} = 1$ when $m_t = i$, 0 otherwise) and R_t the σ-algebra generated by the regime observation. We again use the mean square mode estimate

$$\hat{\phi}_t = E\{\phi_t \mid R_t\}$$

with components $\hat{\phi}_{ti} = \text{Prob}\ \{\phi_t = i \mid R_t\}$, $i = 1$ to N^o.

Of course if r_t is not measured, we have to use the information contained in the plant state. The conflict between regulation objectives and probing actions then defines the dual optimal control problem, as discussed in chapter 7.

To simplify the presentation, the special case of the above example is considered again, that is a system with a single failure ($r_t \in \{1, 2\}$) and the m_t odd/even projection. Denoting by $c_t(i,j)$, $i,j = 1, N^o$, the counting process recording the number of mode transitions from i to j up to time t, the regime measurement is equivalent to the observation of the two counting processes

$$c_t(1) = \sum_{i \in Cl_1} \sum_{j \in Cl_2} c_t(i,j)$$

$$c_t(2) = \sum_{i \in Cl_2} \sum_{j \in Cl_1} c_t(i,j) \tag{8.2.3}$$

Transitions from $r_t = 1$ to $r_t = 2$ (resp. from 2 to 1) are indicated by $\Delta c_t(1) = 1$ (resp. $\Delta c_t(2) = 1$) and $c_t(1)$, $c_t(2)$ obviously have no common jumps. The intensities are

$$\lambda_t(1) = \sum_{i \in Cl_1} \sum_{j \in Cl_2} \pi^o_{ij}\, \phi_{ti} \quad \text{for } c_t(1)$$

$$\lambda_t(2) = \sum_{i \in Cl_2} \sum_{j \in Cl_1} \pi^o_{ij}\, \phi_{ti} \quad \text{for } c_t(2)$$

Estimating the mode from the observations (8.2.3) is a special case of the situation where a Markov chain is to be estimated based on the observation of selected transitions (here odd $\leftrightarrow$ even), studied for example in (Brémaud, 1981). Indeed, theorem 8.1 below could be extracted from (Brémaud, 1981, p. 106), but a specialized formulation and proof are proposed. The initial mode distribution is given by p_i^0 = $\text{Prob}\{m_{t_0} = i\}$.

Theorem 8.1
The mode estimate $\hat{\phi}_t$ is the output of the recursive filter

$$d\hat{\phi}_{ti} = (\Pi^o \hat{\phi}_t)_i \, dt - \hat{\phi}_{ti}^-[dc_t(1) - \sum_{k \in Cl_1} \sum_{j \in Cl_2} \pi_{kj}^o \hat{\phi}_{tk} \, dt]$$

$$- [\hat{\phi}_{ti}^- - (\sum_{j \in Cl_2} \pi_{ji}^o \hat{\phi}_{tj}^-) / (\sum_{j \in Cl_2} \hat{\phi}_{tj}^-) \sum_{k \in Cl_1} \pi_{jk}^o] [dc_t(2) - \sum_{j \in Cl_2} \sum_{k \in Cl_1} \pi_{jk}^o \hat{\phi}_{tj} \, dt] \tag{8.2.4}$$

with $\hat{\phi}_{0i} = p_i^0$

for $i \in Cl_1$ and the symmetric expression for $i \in Cl_2$.

Proof:

It is based on the optimal filter representation theorem and, to apply this result, the problem is imbedded in the following formulation.

It is desired to estimate the indicator ϕ_t given by

$$d\phi_t = \Pi^{o\prime} \phi_t dt + d\mu_t$$

from the counting observations

$$dc_t(1) = \sum_{i \in Cl_1} \sum_{j \in Cl_2} \pi_{ij}^o \phi_{ti} dt + d\mu_t^1$$

$$dc_t(2) = \sum_{i \in Cl_2} \sum_{j \in Cl_1} \pi_{ij}^o \phi_{ti} dt + d\mu_t^2 \tag{8.2.5}$$

where μ_t, μ_t^1 and μ_t^2 are Ξ - martingales ($R_t \subset \Xi_t$). Through the projection (8.2.2) these martingales are not independent. From (Wong and Hajek, 1985) the filter representation is

$$d\hat{\phi}_t = \Pi^{o'}\hat{\phi}_t dt + K_t(1)\,(dc_t(1) - \hat{\lambda}_t(1)\,dt)$$
$$+ K_t(2)\,(dc_t(2) - \hat{\lambda}_t(2)\,dt)$$

where the $\hat{\lambda}_t(\ell)$, $\ell = 1,2$ are the estimated intensities

$$\hat{\lambda}_t(1) = E\{\lambda_t(1) \mid R_t\} = \sum_{i\in Cl_1}\sum_{j\in Cl_2} \pi^o_{ij}\,\hat{\phi}_{ti}$$

and

$$\hat{\lambda}_t(2) = E\{\lambda_t(2) \mid R_t\} = \sum_{i\in Cl_2}\sum_{j\in Cl_1} \pi^o_{ij}\,\hat{\phi}_{ti}$$

The gains $K_t(1)$ and $K_t(2)$ are given by

$$K_t(\ell) = [E\{(\phi_t - \hat{\phi}_t)(\lambda_t(\ell) - \hat{\lambda}_t(\ell))\} + E\{\psi_t(\ell) \mid R_t-\}] \,/\, [E\{\lambda_t(\ell) \mid R_t-\}], \quad \ell = 1,2 \qquad (8.2.6)$$

for

$$\int_{t_0}^{t} \psi_s(\ell)\,ds = <\mu, \mu^\ell>_t \qquad , \ell = 1,2 \qquad (8.2.7)$$

Considering $\ell = 1$, the denominator in (8.2.6) is evaluated as

$$E\{\lambda_t(1) \mid R_t -\} = \sum_{k\in Cl_1}\sum_{j\in Cl_2} \pi^o_{kj}\,\hat{\phi}^-_{tk} \qquad (8.2.8)$$

and the symmetric expression for $\lambda_t(2)$. The first term of the numerator is an N^o-dimensional vector whose i-th component is written as

$$E\{(\phi_{ti} - \hat{\phi}_{ti}) \sum_{k\in Cl_1}\sum_{j\in Cl_2} \pi^o_{kj}\,(\phi_{tk} - \hat{\phi}_{tk}) \mid R_t-\} \qquad (8.2.9)$$

and two cases are distinguished:

- For $i \in Cl_2$, $\phi_{ti}\phi_{tk}$ is zero for any $k \in Cl_1$, so that (8.2.9) reduces to

$$E\{\sum_{k\in Cl_1}\sum_{j\in Cl_2}\pi^o_{kj}(-\phi_{ti}\hat{\phi}_{tk}+\hat{\phi}_{ti}\phi_{tk}+\hat{\phi}_{ti}\hat{\phi}_{tk})\mid R_{t^-}\} = -\hat{\phi}_{ti}^- \sum_{k\in Cl_1}\sum_{j\in Cl_2}\pi^o_{kj}\hat{\phi}_{tk}^- \quad (8.2.10a)$$

- For $i \in Cl_1$, $\phi_{ti}\phi_{tk}$ is zero for any $k \in Cl_1$ except when $k = i$ and it is then equal to ϕ_{ti}^2 ($=\phi_{ti}$), so that (8.2.9) reduces to

$$-\hat{\phi}_{ti}^- \sum_{k\in Cl_1}\sum_{j\in Cl_2}\pi^o_{kj}\hat{\phi}_{tk}^- + \hat{\phi}_{ti}^- \sum_{j\in Cl_2}\pi^o_{ij} \quad (8.2.10b)$$

The $\psi_t(1)$ term is also an N° dimensional vector, and, denoting by $\psi_t(1,i)$ its ith component, one has to evaluate, from (8.2.5), (8.2.7)

$$E\{d\phi_{ti}dc_t(1) \mid R_{t^-}\} = E\{\psi_t(1,i) \mid R_{t^-}\}\, dt$$

Again separating i odd and i even:

- For $i \in Cl_1$, $d\phi_{ti}\, dc_t(1)$ is zero except when $d\phi_{ti} = -1$ (jump out of Cl_1 from i) and $dc_t(1)$ is then equal to $\sum_{j\in Cl_2}\pi^o_{ij}\phi_{ti}\,dt$, plus the martingale term, so that $E\{\psi_t(1,i) \mid R_{t^-}\}$ reduces to

$$-\sum_{j\in Cl_2}\pi^o_{ij}\hat{\phi}_{ti}^- \quad (8.2.11a)$$

- For $i \in Cl_2$, $d\phi_{ti}\, dc_t(1)$ is zero except when $d\phi_{ti} = 1$ (jump into Cl_2 at i) and the contributing part of $dc_t(1)$ is then equal to $\sum_{k\in Cl_1}\pi^o_{ki}\phi_{tk}\,dt$, so that (8.2.11a) is replaced by

$$+\sum_{k\in Cl_1}\pi^o_{ki}\hat{\phi}_{tk}^- \quad (8.2.11b)$$

Expressions symmetric to (8.2.9), (8.2.10a,b) and (8.2.11a,b) are obtained for $\ell = 2$ and, collecting all this finally produces (8.2.4) of the theorem and its counterpart for $i \in Cl_2$. □

For implementation, it might be interesting to look for a linear version of (8.2.4) with the unnormalized probabilities, as explained in chapter 6.

8.3 Wide-Band Hybrid Models

To design the steam temperature regulator for the solar thermal receiver presented in chapter 1, a coarse lumped linear approximation had been confirmed by some frequency response tests on an actual panel. However, the low-pass character of the transfer function

$$C(sI - A)^{-1}B$$

associated with

$$\dot{x} = Ax + Bu, \qquad y = Cx$$

is not satisfied for this application. The reason is that each section of the lumped receiver model gives rise to a pole and zero pair and the complete transfer function is only proper. There is no doubt that tests at higher frequencies would reveal that the actual receiver dynamics cut really fast signals but, within the range of interest, it appears that we have to accept an unusual wide-band behaviour and tailor our state-space model to it. We decompose the observation y_t into a direct control to output channel (wide-band) and the output of a linear time-invariant (low-pass) model with

$$x_t = \xi_t + D\,u_t$$

$$\dot{\xi}_t = A\,\xi_t + B\,u_t \tag{8.3.1}$$

The transfer is then wide-band $X(s) = [(sI - A)^{-1}B + D]\,U(s)$. Of course we are still interested in hybrid dynamics and the parameters A, B, D are indexed by the regime $r_t \in S = \{1, 2, ..., N\}$. We shall assume that r_t transitions are markovian with generator Π'. Remember that for the solar receiver, changes in r_t reflect sudden jumps of the insolation as a cloud crosses over the field of heliostats.

We could recover the low-pass character by simply adding high frequency poles to the model. This would, however, unnecessarily increase the order and complicate the design. It would also be extremely difficult to determine experimentally the location of these additional poles and the corresponding new states would be hard to track using existing sensors. On the other hand, as we shall explain below, it is not too complicated to extend the design of chapter 3 to a wide-band model like (8.3.1).

We consider again an average quadratic cost

$$J = E\{ \int_{t_0}^{t_f} (x_t'Q_1x_t + u_t'Q_2u_t)dt \mid x_0, i_0, t_0\} \tag{8.3.2}$$

where the symmetric weighting matrices ($Q_1 \geq 0$, $Q_2 > 0$) can be made regime dependent ($Q_1 = Q_{1i}$, $Q_2 = Q_{2i}$ when r_t is i). It is assumed that the regime is measured and the information available at the regulator is X_t v R_t with $R_t = \sigma$ - $\{r_s, t_0 \leq s \leq t\}$, $x_t = \sigma$ - $\{x_s, t_0 \leq s \leq t\}$ for x_t given by (8.3.1). The class of admissible control laws $\mathscr{Z}$ thus consists of X_t v R_t measurable functions satisfying usual smoothness conditions (see appendix 2).

At the instant of a jump, we must remember that the plant equations are non linear and that (8.3.1) is only a linearized approximation, valid in a limited neighbourhood of the nominal level $x_n(i)$ when $r_t = i$. When the $i \rightarrow j$ regime transition occurs, we have

$$\xi_t^- = x_t^- - D(i)\, u_t^-$$

$$\xi_t = x_t - D(j)\, u_t$$

and a discontinuity

$$\xi_t = \xi_t^- + \delta(j,i) + D(i)\, u_t^- - D(j)\, u_t \tag{8.3.3}$$

where $\delta(j,i) = x_n(i) - x_n(j)$. The jump condition displayed by (8.3.3) is different from the condition used in chapter 7 and is rather unusual with a control dependent character. It directly results from the wide-band influence and if the D_is are set to zero, we recover a discontinuity such as that of chapter 7.

With the model (8.3.1), (8.3.3) for the state and the regime markov chain, the optimal regulator is defined as

$$u^* = \operatorname*{Arg\,min}_{u \in \mathscr{Z}} J \tag{8.3.4}$$

Optimization is carried out explicitly and we obtain

<u>Theorem 8.2</u>

The solution of (8.3.4) is, assuming that $I + \bar{\Gamma}_i D_i$ is invertible for i = 1 to N,

$$u_t^* = (I + \bar{\Gamma}_i D_i)^{-1} (\bar{\Gamma}_i x_t + \bar{\Delta}_i) \quad \text{when } r_t = i \tag{8.3.5}$$

with the notations

$$\begin{aligned} \bar{\Gamma}_i &= \Gamma_i + D_i' \sum_{j=1}^{N} \pi_{ij} \Lambda_j D_j \Gamma_j \\ & \qquad\qquad\qquad\qquad , \ i = 1, \ N \\ \bar{\Delta}_i &= \Delta_i + D_i' \sum_{j=1}^{N} \pi_{ij} \Lambda_j D_j \Delta_j \end{aligned} \tag{8.3.6}$$

where

$$\Gamma_i = - \bar{Q}_{2i}^{-1} (B_i' \Lambda_i + D_i'(Q_1 + \sum_{j=1}^{N} \pi_{ij} \Lambda_j))$$

$$\Delta_i = - \bar{Q}_{2i}^{-1} (B_i' \lambda_i + D_i' \sum_{j=1}^{N} \pi_{ij} (\lambda_j + \Lambda_j \delta(j, i))$$

$$\bar{Q}_{2i} = Q_2 + D_i'(Q_1 + \sum_{j=1}^{N} \pi_{ij} \Lambda_j) D_i$$

The matrices Λ_i, i = 1, N, and vectors λ_i, i = 1, N, are given by the equations

$$\begin{aligned} - \dot{\Lambda}_i &= \Lambda_i(A_i + B_i\bar{\Gamma}_i) + (A_i + B_i\bar{\Gamma}_i)'\Lambda_i + \bar{\Gamma}_i'\bar{Q}_{2i}\bar{\Gamma}_i + (Q_1 + \sum_{j=1}^{N} \pi_{ij} \Lambda_j)(I/2 + D_i\bar{\Gamma}_i) \\ &+ (I/2 + D_i\bar{\Gamma}_i)'(Q_1 + \sum_{j=1}^{N} \pi_{ij} \Lambda_j) \\ &+ \sum_{j=1}^{N} \pi_{ij} (\bar{\Gamma}_i'D_j'\Lambda_jD_j\bar{\Gamma}_j - (I + \bar{\Gamma}_i'D_i')\Lambda_jD_j\bar{\Gamma}_j - \bar{\Gamma}_j'D_j'\Lambda_j(I + \bar{\Gamma}_i'D_i)) \end{aligned} \tag{8.3.7}$$

$\Lambda_i(t_f) = 0$; i = 1, N

$$
\begin{aligned}
-\dot{\lambda}_i = {} & (A_i + B_i\bar{\Gamma}_i)\lambda_i + \Lambda_i B_i\bar{\Delta}_i + \frac{1}{2}(\bar{\Gamma}_i'Q_2\bar{\Delta}_i + \bar{\Delta}_i'Q_2\bar{\Gamma}_i) \\
& + \frac{1}{2}((I + D_i\bar{\Gamma}_i)'Q_1D_i\bar{\Delta}_i + \bar{\Delta}_i'D_iQ_1(I + D_i\bar{\Gamma}_i)) + \sum_{j=1}^{N} \pi_{ij}(I + D_i\bar{\Gamma}_i - D_j\bar{\Gamma}_j)\lambda_j \qquad (8.3.8) \\
& + \frac{1}{2}(\sum_{j=1}^{N} \pi_{ij}(I + D_i\bar{\Gamma}_i - D_j\bar{\Gamma}_j)'\lambda_j(\delta(j,i) + D_i\bar{\Delta}_i - D_j\bar{\Delta}_j) \\
& + \sum_{j=1}^{N} \pi_{ij}(\delta(j,i) + D_i\bar{\Delta}_i - D_j\bar{\Delta}_j)'\lambda_j(I + D_i\bar{\Gamma}_i - D_j\bar{\Gamma}_j))
\end{aligned}
$$

$\lambda_i(t_f) = 0;\ i = 1, N$

Proof:

The optimal cost-to-go could be parametrized by x as we have done in previous chapters, but here it is more convenient to use ξ and define

$$
V(\xi, i, t) = \min_{\mathscr{U}} E\{ \int_t^{t_f} ((\xi_s + Du_s)'\, Q_1(\xi_s + Du_s) + u_s'Q_2u_s)ds \mid \xi_{s^-} = \xi,\ r_{t^-} = i,\ t\}
$$

The verification theorem of dynamic programming leads to

$$
0 = \min_{u \in \mathbf{R}^m} \{V_t + \mathscr{L}_u V + x_t'\, Q_1 x_t + u_t'Q_2u_t\}
$$

For the model (8.3.1) we find

$$
\mathscr{L}_u V(\xi, i, t) = V_x'(A_i\xi_t + B_iu_t) + \sum_{j=1}^{N} \pi_{ij}\, E\{V(\xi_t, j, t) \mid \xi_{t^-} = \xi,\ r_{t^-} = i,\ r_t = j\}
$$

If we assume that V is quadratic in ξ with coefficients indexed by the regime,

$$
V(\xi, i, t) = \xi'\, \Lambda_i\xi + 2\lambda_i'\xi + \mu_i
$$

the terms involving u in the above minimization are, for $r_{t^-} = i$,

$$(2\Lambda_i \xi_{t^-} + 2\lambda_i)' (A_i\xi_{t^-} + B_iu_{t^-})$$

$$+ \sum_{j=1}^{N} \pi_{ij}(\xi_{t^-} + \delta(j, i) + D_iu_{t^-} - D_ju_t)' \Lambda_j(\xi_{t^-} + \delta(j, i) + D_iu_{t^-} - D_ju_t)$$

$$+ \sum_{j=1}^{N} 2\pi_{ij}\lambda_j'(\xi_{t^-} + \delta(j, i) + D_iu_{t^-} - D_ju_t)$$

$$+ (\xi_{t^-} + D_iu_{t^-})' Q_1(\xi_{t^-} + D_iu_{t^-}) + u_{t^-} Q_2u_{t^-}$$

But there are no constraints and the minimum is obtained by setting the partial derivative w.r.t. u to zero

$$0 = 2Q_2u_{t^-} + 2D_i'Q_1D_iu_{t^-} + 2(\sum_{j=1}^{N} \pi_{ij} D_i'\Lambda_jD_i) u_{t^-} + 2B_i'\Lambda_i\xi_{t^-} + 2B_i'\lambda_i$$
$$+ 2D_i'Q_1\xi_{t^-} + 2D_i' \sum_{j=1}^{N} \pi_{ij} \lambda_j + 2D_i' \sum_{j=1}^{N} \pi_{ij} \Lambda_j(\xi_{t^-} + \delta(j, i) - D_ju_t)$$

Using the notations of the theorem, this is written

$$u_{t^-} = \Gamma_i\xi_t + \Delta_i + D_i' \sum_{j=1}^{N} \pi_{ij} \Lambda_j D_ju_t$$

where $u_{t^-} = u(r_{t^-} = i)$ and $u_t = u(r_t = j)$. Solving for the general expression of u, we obtain

$$u_t = (\Gamma_i + D_i' \sum_{j=1}^{N} \pi_{ij} \Lambda_j D_j\Gamma_j) \xi_t + \Delta_i + D_i' \sum_{j=1}^{N} \pi_{ij}\Lambda_j D_j\Delta_j$$

or, with $\bar{\Gamma}_i$ and $\bar{\Delta}_i$, i = 1, N, defined by (8.3.6)

$$u_t = \bar{\Gamma}_i \xi_t + \bar{\Delta}_i$$

However, ξ_t is not directly observed and we need to write u as a function of x_t. Using

$$\xi_t = x_t - D(r_t) u_t$$

and assuming that $I + \bar{\Gamma}_i D_i$ is invertible for i = 1 to N, we find

$$u_t = (I + \bar{\Gamma}_i D_i)^{-1} (\bar{\Gamma}_i x_t + \bar{\Delta}_i) \quad \text{when } r_t = i$$

which proves (8.3.5). Bringing this expression back into the optimal cost-to-go, we get a quadratic function equal to $-(\xi'\dot{\Lambda}_i\xi + 2\dot{\lambda}_i'\xi + \dot{\mu}_i)$. Identifying the coefficient of $\xi'\xi$ to $-\dot{\Lambda}_i$ leads to

$$
\begin{aligned}
-\dot{\Lambda}_i = {} & \Lambda_i(A_i + B_i\bar{\Gamma}_i) + (A_i + B_i\bar{\Gamma}_i)'\Lambda_i + (Q_1 + \sum_{j=1}^{N} \pi_{ij} \Lambda_j)(I + D_i\bar{\Gamma}_i) \\
& + \bar{\Gamma}_i'D_i'(Q_1 + \sum_{j=1}^{N} \pi_{ij} \Lambda_j) + \bar{\Gamma}_i'(Q_2 + D_i'Q_1D_i + \sum_{j=1}^{N} \pi_{ij} D_i'\Lambda_jD_i)\,\bar{\Gamma}_i' \\
& + \sum_{j=1}^{N} \pi_{ij} (\bar{\Gamma}_j'D_j'\Lambda_jD_j\bar{\Gamma}_j - \Lambda_jD_j\bar{\Gamma}_j - \bar{\Gamma}_j'D_j'\Lambda_j - \bar{\Gamma}_i'D_i'\Lambda_jD_j\bar{\Gamma}_j - \bar{\Gamma}_j'D_j'\Lambda_jD_i\bar{\Gamma}_i)
\end{aligned}
$$

which is reorganized to obtain (8.3.7) of the theorem, with the boundary condition $\Lambda_i(t_f) = 0$, i = 1, N, because the cost-to-go vanishes at $t = t_f$.

The constant term would give the $-\dot{\mu}_i$ equation but this does not influence the optimal control and we do not give it. Finally the linear term produces, after lengthy but direct manipulations, equation (8.3.8) for the vectors λ_i with terminal condition $\lambda_i(t_f) = 0$, i = 1 to N. □

The design equations (8.3.7), (8.3.8) are not unlike equations encountered previously. In particular (8.3.7) is akin to the set of coupled Riccati equations of chapter 3, but, because of the dependence of $\bar{\Gamma}_i$ on $\bar{Q}_{2i} = \bar{Q}_{2i}\,(\Lambda_j, j = 1, N)$ it is rational rather than quadratic in its unknowns. This situation was also encountered in chapter 5 in the presence of multiplicative noises. Once the Λ_is are determined, (8.3.8) is simply a set of coupled linear vector differential equations in the λ_is.

As in chapter 7, the optimal regulator (8.3.5) has two terms, a linear feedback proportional to the error variable and a bias. This bias results from the change of set point when the regime jumps and indeed, if $\delta(i, j)$ is set to zero for i, j = 1, N it is easy to check from (8.3.6), (8.3.8) that $\lambda_i = 0$, i = 1, N, and u_t^* reduces to its feedback term ($\bar{\Delta}_i = 0$, i = 1, N). Since they do not depend on δ the Λ_is remain unchanged.

The computation of the regulator of theorem 8.2 is involved with the need to integrate simultaneously coupled matrix or vector differential equations. Going back to the solar receiver example, we find some motivation for a useful approximation. Even though the boiler dynamics revealed a wide-band character, this phenomenon remains small, although not negligible. If this were not the case it would indicate the need for further testing to increase the bandwidth over which the model provides a satisfactory description of the plant. Mathematically this translates as a smallness of the D_i, $i = 1, N$, matrices and it is possible to approximate equations (8.3.5) to (8.3.6) by retaining only first-order terms in D. We obtain

$$u_t = (I + \Gamma_i D_i)^{-1} (\Gamma_i x_t + \Delta_i) \quad \text{when } r_t = i$$

with

$$\Gamma_i = - Q_2^{-1} (B_i'\Lambda_i + D_i'(Q_1 + \sum_{j=1}^{N} \pi_{ij} \Lambda_j))$$

$$\Delta_i = - Q_2^{-1} (B_i'\lambda_i + D_i'(Q_1 + \sum_{j=1}^{N} \pi_{ij} (\lambda_j + \Lambda_j\delta(j, i)))$$

and the Λ_is given as solutions of the simplified equations

$$- \dot{\Lambda}_i = A_i'\Lambda_i + \Lambda_i A_i - \Lambda_i B_i Q_2^{-1} B_i'\Lambda_i$$
$$+ (Q_1 + \sum_{j=1}^{N} \pi_{ij} \Lambda_j)(I/2 - D_i Q_2^{-1} B_i'\Lambda_i) + (I/2 - D_i Q_2^{-1} B_i'\Lambda_i)'(Q_1 + \sum_{j=1}^{N} \pi_{ij} \Lambda_j)$$
$$+ \sum_{j=1}^{N} \pi_{ij}(\Lambda_j D_j Q_2^{-1} B_j'\Lambda_j + \Lambda_j B_j Q_2^{-1} D_j'\Lambda_j)$$

Similarly, simpler equations are obtained for the λ_is.

8.4 Multiple Decision Makers

Modern game theory was born about half a century ago, around economic planning and forecasting applications: the thirties' economic crisis and World War II induced a global reorganization of industrial and financial activities. With the growth of multinational companies the need for strategic planning appeared and this motivated research in models; static at first and then dynamic, of economic agents (companies, trusts, governmental agencies, etc.) competing to dominate or regulate the market.

The historical reference is the seminal book by Von Neumann and Morgenstern (1944), following work in Germany by Von Stackelberg (1934), less accessible at the time. In fact we already touched upon this theory in chapter 4 when we interpreted stochastic control as a game between the designer and Nature. The strategy there was defined as the way the designer elects to cope with the variability introduced by Nature's choices, and we studied mean optimal, equalizing and minimax strategies. In the general context the strategy also corresponds to a relationship between the players.

This is typically illustrated by another well-known application of game theory in military pursuit and evasion problems. When an interceptor (aircraft, missile, etc.) tracks a target (aircraft, ship, etc.) its objective is to minimize a distance measure while the target tries to evade the track by pursuing the opposite objective. The relationship to optimization is apparent and indeed (Isaac, 1965), based on results anterior to Pontryagin's and Bellman's publications, shows how game theory generalizes optimal control to the case of several decision makers.

For process control applications, game aspects appear with hierarchical and distributed control structures. To solve problems of increasing complexity (large chemical plants, space stations, networks of surveillance systems, etc.) it is interesting to perform part of the data processing at a local level and to transfer only an aggregate decision problem to a centralized upper-level unit. The relationship between the various (local or central) decision makers is then phrased in terms of game strategies: who knows what and when, what authority is allowed here and there, etc. We shall restrict our attention to two concepts of game theory which are immediately relevant, Nash and Stackelberg strategies.

Consider a game with two players, each with a performance objective J_i and a decision S_i, i = 1, 2. The open-loop Nash equilibrium gives optimal choices S_i^*, i = 1, 2, satisfying

$$\begin{aligned} J_1(S_1^*, S_2^*) &\le J_1(S_1, S_2^*) \\ J_2(S_1^*, S_2^*) &\le J_2(S_1^*, S_2) \end{aligned} \qquad \forall\ S_1,\ S_2$$

In other words, the two players have equal authority and no one can improve his performance by modifying his decision alone. This strategy describes the completely decentralized situation in large-scale systems where no on-line communication between local units is allowed.

The second strategy corresponds to a hierarchical control situation where some supervisor unit assumes leadership for the other units, the Stackelberg strategy. In a Stackelberg game, one player, say the second, is the leader and declares his decision first. The other player, called the follower, then makes a decision taking into account the leader's declaration. Assuming that for a fixed S_2 there is a unique S_1, with $S_1 = f(S_2)$, Stackelberg's equilibrium is defined by S_1^* and S_2^* such that

$$\text{leader} \quad J_2(f(S_2^*), S_2^*) \leq J_2(f(S_2), S_2) \quad \forall\, S_2$$
$$\text{follower} \quad J_1(f(S_2^*), S_2^*) \leq J_1(f(S_1), S_2) \quad \forall\, S_2 \text{ and } S_1$$

For the leader, playing a Stackelberg strategy ensures a return at least as good as its return under a Nash strategy.

In some applications it is interesting to model the dynamics underlying the game by a hybrid system. Through the random jumps of the regime this introduces an additional player, Nature, as we interpreted it in chapter 4, deciding upon the actual values of the state equation parameters. More interestingly the strategy can also be made contingent upon the regime, reflecting random changes in the relationship between players. We shall consider a typical game with three regimes and two players where r = 1 gives a Stackelberg strategy with player 1 as the leader, r = 2 also a Stackelberg strategy but with player 2 as the leader, and r = 3 a Nash strategy. Regime transitions then describe the fact that leadership is often a changing "gift of Nature" and hence may rotate among players. There may also be instances of the random environment where no player is strong enough to be declared a leader, leading to a Nash equilibrium.

We now formalize the above problem statement in mathematical terms. The state dynamics are influenced by two control signals u_{p1} and u_{p2} decided by player 1 and 2 respectively,

$$\dot{x}_{pt} = f(x_{pt}, u_{p1t}, u_{p2t}, \phi_t) \tag{8.4.1}$$

where ϕ_t, the regime indicator, is given by a Markov chain as before. We take three regimes as explained above ($S = \{1, 2, 3\}$) so that the chain generator is a 3 x 3 matrix Π'. Player k minimizes a cost

$$J_k = E\{ \int_{t_0}^{t_f} g_k(x_{pt}, u_{p1t}, u_{p2t}, \phi_t) \mid x_{t_0}, r_{t_0}, t_0\} \qquad k = 1, 2 \tag{8.4.2}$$

Admissible control laws, $\mathscr{Z}_1$ and $\mathscr{Z}_2$, are constrained as usual by smoothness conditions but in a game context the information structure assigned to the two players must also be defined very precisely. For each strategy, the equilibrium is in general different when the game is played in open-loop, closed-loop or feedback forms. A discussion of these concepts and the associated non uniqueness of the equilibrium may be found in (Basar and Olsder, 1982) but here we restrict our attention to feedback information structures: we assume that x_{pt} and r_t are exactly measured and the players' decisions are smooth direct feedback of these quantities $u_{p1t}(x_{pt}, r_t)$ and $u_{p2}(x_{pt}, r_t)$. Note that this is the structure used in the single decision maker situation in chapter 3. We shall often indicate the current regime by an index and, to avoid confusion, we shall reserve index k (k = 1, 2) for the players and index i (i = 1, 2, 3) for the regimes.

Using dynamic programming, the first step of the optimization is to obtain the Hamilton-Jacobi-Bellman equation for the cost-to-go. We use the parametrization $V_k(x_p, r, t)$ for k = 1, 2. Assuming that the game has an equilibrium under admissible strategies, $u_1^* \in \mathscr{Z}_1$ and $u_2^* \in \mathscr{Z}_2$, the cost-to-go satisfies

$$\frac{\partial}{\partial t} V_k(t, x_p, i) + f'(x_p, u_{p1}^*, u_{p2}^*, i) \frac{\partial}{\partial x} V_k(x_p, i, t) + \sum_{j=1}^{3} \pi_{ij} V_k(t, x_p, j)$$
$$+ g_k(x_p, u_{p1}^*, u_{p2}^*, i) = 0 \quad ; \text{when } r_t = i, \quad \text{for } k = 1, 2 \tag{8.4.3}$$

with $V_k(x_p, i, t_f) = 0, \quad i = 1, 3.$

In general the solution of (8.4.3) is difficult and one cannot extract the optimizing control

$$u_{pk}^* = \operatorname*{Arg\,min}_{u_k \in \mathscr{Z}_k} \{f'(x_p, u_{p1}, u_{p2}, i) \frac{\partial}{\partial x} V_k(x_p, i, t) + g_k(x_p, u_{p1}, u_{p2}, i)\} \tag{8.4.4}$$

when $r_t = i$, for k = 1, 2.

The model is therefore specialized to linear state dynamics. As explained in chapters 3 and 7 we then use x instead of x_p (and u instead of u_p) to signal the underlying linearization procedure

$$\dot{x}_t = A_i\, x_t + B_{1i}\, u_{1t} + B_{2i}\, u_{2t} \tag{8.4.5}$$

Only quadratic costs are considered

$$g_k = x_t'\, Q_{1k}\, x_t + u_{kt}'u_{kt} - 2u_{kt}'\, Q_{2k\underline{k}} u_{\underline{k}t} \tag{8.4.6}$$

where $\underline{k}$ designates the other player ($\underline{1} = 2$, $\underline{2} = 1$).

We then have

<u>Theorem 8.3</u>

Under the above assumptions, the solution (8.4.4) of the JLQ game (8.4.5), (8.4.6) is given by

$$u^*_{kt} = -\,(\underline{\mathcal{Q}}_{kki}\, \Lambda_{ki} + \underline{\mathcal{Q}}_{k\underline{k}i}\, \Lambda_{\underline{k}i})\, x_t \qquad \text{when } r_t = i \tag{8.4.7}$$

with matrices

$$\underline{\mathcal{Q}}_{kki} = \begin{cases} -\,(I - Q_{2k\underline{k}}\, Q_{2\underline{k}k} - Q_{2\underline{k}k}'Q_{2k\underline{k}}')^{-1}\, (B'_{ki} + Q_{2\underline{k}k}'\, B'_{\underline{k}i}) & \text{for } i = k \\ Q_{2k\underline{k}}\, \underline{\mathcal{Q}}_{\underline{k}k\underline{k}} - B'_{ki} & \text{for } i = \underline{k} \\ -\,(I - Q_{2k\underline{k}}\, Q_{2\underline{k}k})^{-1}\, B'_{k3} & \text{for } i = 3 \end{cases} \tag{8.4.8}$$

$$\underline{\mathcal{Q}}_{k\underline{k}i} = \begin{cases} -\,(I - Q_{2k\underline{k}}\, Q_{2\underline{k}k} - Q_{2\underline{k}k}'Q_{2k\underline{k}}')^{-1}\, Q_{2k\underline{k}}'\, B'_{\underline{k}i} & \text{for } i = k \\ Q_{2k\underline{k}}\, \underline{\mathcal{Q}}_{\underline{k}\underline{k}\underline{k}} & \text{for } i = \underline{k} \\ -\,(I - Q_{2k\underline{k}}\, Q_{2\underline{k}k})^{-1}\, B'_{\underline{k}3} & \text{for } i = 3 \end{cases} \tag{8.4.9}$$

and Λ_{ki}, k = 1, 2, i = 1, 3, solutions of a set of coupled equations

$$
\begin{aligned}
-\dot{\Lambda}_{ki} = {} & \Lambda_{ki}\underline{A}_i + \underline{A}_i\Lambda_{ki} + \sum_{j=1}^{3}\pi_{ij}\,\Lambda_{kj} + Q_{1k} \\
& + (\underline{Q}_{2kki}\Lambda_{ki} + \underline{Q}_{2k\underline{k}i}\Lambda_{\underline{k}i})'(\underline{Q}_{2kki}\Lambda_{ki} + \underline{Q}_{2k\underline{k}i}\Lambda_{\underline{k}i}) \\
& - (\underline{Q}_{2kki}\Lambda_{ki} + \underline{Q}_{2k\underline{k}i}\Lambda_{\underline{k}i})'\,Q_{2k\underline{k}}\,(Q_{2\underline{k}ki}\Lambda_{ki} + \underline{Q}_{2\underline{k}\underline{k}i}\Lambda_{\underline{k}i}) \qquad (8.4.10) \\
& - (\underline{Q}_{2\underline{k}ki}\Lambda_{ki} + \underline{Q}_{2\underline{k}\underline{k}i}\Lambda_{\underline{k}i})'\,Q_{2k\underline{k}}'\,(Q_{2kki}\Lambda_{ki} + \underline{Q}_{2k\underline{k}i}\Lambda_{\underline{k}i})
\end{aligned}
$$

$$\Lambda_{ki}(t_f) = 0$$

where $\underline{A}_i = A_i + B_{1i}\Lambda_{1i} + B_{2i}\Lambda_{2i}$.

Proof:

We consider first the Nash case i = 3. Both players are then equal and the minimization in (8.4.4) is readily performed. For a quadratic form of the cost-to-go

$$V_k(x, i, t) = x'\Lambda_{ki}\,x$$

we obtain at the minimum

$$u^*_{kt} - Q_{2k\underline{k}}\,u^*_{\underline{k}t} + B_{ki}'\Lambda_{ki}\,x = 0 \qquad \text{when } i = 3$$
$$k = 1, 2$$

Solving by substitution, we find an equation of u^*_{kt}

$$u^*_{kt} - Q_{2k\underline{k}}\,(Q_{2\underline{k}k}\,u^*_{kt} - B_{\underline{k}i}'\Lambda_{\underline{k}i}\,x) + B_{ki}'\Lambda_{ki}\,x = 0 \qquad \text{when } i = 3$$
$$k = 1, 2$$

which is solved as

$$u^*_{kt} = -(I - Q_{2k\underline{k}}\,Q_{2\underline{k}k})^{-1}(B_{ki}'\Lambda_{ki} + Q_{2k\underline{k}}\,B_{\underline{k}i}'\Lambda_{\underline{k}i})\,x \qquad \text{when } i = 3$$

This proves (8.4.7), (8.4.8), (8.4.9) for i = 3. Next consider i = 1 or 2, i.e. the Stackelberg case. When player k is the leader, the first step is to compute how player $\underline{k}$ follows on, that is to solve

$$\operatorname*{Arg\,min}_{u_{\underline{k}}} \{(Ax + B_{1i}u_1 + B_{2i}u_2)' \frac{\partial}{\partial x} V_{\underline{k}}(x, i, t) + x'Q_{1\underline{k}}\, x + u_{\underline{k}t}\, u_{\underline{k}t}' - u_{\underline{k}t}'\, Q_{2\underline{k}k}\, u_{kt}\}$$

At the minimum we get the answer of player $\underline{k}$ equal to $Q_{2\underline{k}k}\, u_{kt} + B_{\underline{k}i}'\, \Lambda_{\underline{k}i}\, x_t$ for any admissible u_{kt}. Bringing this back into (8.4.4) gives us an optimization for the independent variable, u_{kt}, and it is solved as above

$$0 = u_{kt} - Q_{2k\underline{k}}\, Q_{2\underline{k}k}\, u_{kt} - Q_{2\underline{k}k}'\, Q_{2k\underline{k}}'\, u_{kt} - Q_{2k\underline{k}}\, B_{\underline{k}i}'\, \Lambda_{\underline{k}i}\, x_t$$

$$+ B_{ki}'\, \Lambda_{ki}\, x_t + Q_{2\underline{k}k}\, B_{\underline{k}i}'\, \Lambda_{ki}\, x_t$$

Using $u^*_{\underline{k}t} = Q_{2\underline{k}k}\, u^*_{kt} + B_{ki}'\, \Lambda_{ki}\, x_t$, we obtain (8.4.7) for the notations (8.4.8), (8.4.9), i = k, i = $\underline{k}$. The last part of the proof is to bring back the above optimal decisions into (8.4.3). This gives a quadratic form in x and its nullity for arbitrary x finally gives (8.4.10) with boundary conditions from the definition of V. □

The game of theorem 8.3 describes the situation where Nature decides on the hierarchy, through regime jumps. Players consider a performance index averaged over the possible choices of Nature. In terms of economic planning this is not a completely satisfactory strategy and a wise manager would try to secure his business against "far from average" situations. An interesting research direction is therefore to bring together the robustness ideas in chapter 4 and the above discussion to arrive at equalizing or minimax strategies with a changing hierarchy of players.

8.5 Notes and References

We have presented three possible extensions of the basic hybrid models and illustrated the impact of these modifications by solving control and filtering problems. At the expense of a little increase in complexity, most of the designs in previous chapters carry over to the more general model.

Some references for the non markov, or more precisely semi-markov, case were given in §8.2. We only recall the work of (Sworder, 1980) and a recent application to tracking (Mookerjee et al., 1987), see also (Chizeck and Lalonde, 1984). Non markovian transitions still cause many significant difficulties and there exists a rich field of research for hybrid processes where the discrete variable is a renewal process. Initial results on this arduous problem are reported in (Malhame, 1988), motivated by a heating/cooling large-scale regulation problem, see also (Sworder and Elliott, 1986). The introduction given in §8.2 closely follows (Mariton, 1988).

On the multiple decision makers problem we relied on (Basar and Haurie, 1984), which contains many other ideas for further study. The presentation of wide-band hybrid models in §8.3 evolved from (Sworder, 1982). The difference is that we used a discontinuity model such as that in chapter 7 whereas in Sworder's paper the discontinuities are as in chapter 5 with Poisson impulsive disturbances.

When a machine is used intensively in a difficult environment, its life expectation decreases. Also, when an actuator stays for a long interval stuck at its saturation value its failure rate is increased. This suggests that we should sometimes consider state and control dependent transition rates, $\Pi = \Pi(x, u)$. The problem was analyzed in (Sworder and Robinson, 1973, Robinson and Sworder, 1974). The control u_t influences the state trajectories directly through the plant state dynamics and also through $\Pi(x, u)$ where it can modify the regime transition rates. Design of a regulator in such a situation is very difficult and simplifications must be made. The dependence of Π on x and u can be expanded in terms of a small parameter as $\pi_{ij} = \pi_{ij}^0(1 + \varepsilon(\alpha'_i\, x + \beta'_i\, u) + o(\varepsilon))$. For $\varepsilon = 0$ our usual model is recovered and for ε small a first-order approximation of the optimal regulator (w.r.t. a quadratic cost) is presented in (Sworder and Robinson, 1973). Examples indicate that the near optimal regulator uses its influence on the regime jumps in a rather subtle way: of course it has a tendency to favor transition to more favorable regimes (smaller cost penalties, more stable open loop dynamics, etc.) by increasing the corresponding rates, but this has the rather unexpected consequence that the regulator sometimes decides to apply unnecessarily large control amplitudes, just to make sure that the resulting state trajectories facilitate a transition to favored regimes. There are situations, however, where the weak dependence assumption cannot be made. For this case (Robinson and Sworder, 1974) propose to restrict attention to the class of hybrid systems with a directed chain ($\pi_{ii-1} = -\pi_{ii}$, $i = 2, N$, $\pi_{11} = 0$) and, under this assumption, they obtain the optimal solution in open-loop form via the iterative solution of a nonlinear two-point boundary value problem. With the language of point processes theory the situation where u influences Π is called intensity control. The controller can modulate the intensity of regime transitions but cannot add or erase a transition. There are applications, however, where we have the flexibility to command a transition. In a redundant fault-tolerant system, for example, we can decide the instants when operational components are turned on and off. An interesting control problem is to optimize this scheduling on the basis of performance and reliability. Future research on hybrid systems should consider this aspect. The dynamic programming conditions for optimal impulse control are in general different from the usual Hamilton-Jacobi-Bellman but we can use the theory of quasivariational inequalities developed by (Bensoussan and Lions, 1975).

The dual control problem also still requires much research effort. All we have at present in the literature are a few heuristic suboptimal solutions and we are far from a systematic methodology to include dual considerations in a given application. The discussion in §7.6 and its references provide a starting point for future research.

In our work we have never discussed directly the dimension of the model, i.e. the dimension n of the state space and the cardinal N of the regime set. If they are very large, say n and N larger than 50, it becomes very difficult to analyze the model globally. Filter or regulator design is also complicated. Within this large number of variables some may be fast and others slow, and a simpler model could result by neglecting fast dynamics on long time intervals. This approach is well known as perturbation analysis in the non hybrid case but there remain open questions regarding its applicability when fast and slow dynamics can be both state components or regime transitions. This is the subject of on-going research and representative available results are (Castanon et al., 1980, Ezzine and Johnson, 1986, Ezzine and Haddad, 1988).

Communication Command and Control (C3) applications illustrate the dimensionality problem, the best example being large military BM/C3 (Battle Management) systems like the SDI (Strategic Defense Initiative). Related civilian applications include air traffic control, automated transportation systems and nuclear reactors. The state space to describe these systems is naturally hybrid with euclidean variables representing typically the positions of missiles, weapons, radar, satellites, etc., and discrete variables indexing the type of the missile, its friend or foe classification, resource availability status, etc. In the position paper (Athans, 1987) the development of a comprehensive theory for hybrid dynamic processes is listed as a key step towards understanding the complexities of C3 applications. Basically all the issues of C3 involve hybrid control and estimation algorithms and it is hoped that this book will be useful in this context. The future of research on hybrid systems will certainly be largely influenced by the needs of these applications. Directions of immediate interest are the modelling of human decision makers in an uncertain BM/C3 environment, distributed multiple target tracking and decentralized resource allocations.

References

Athans, M. (1987). Command and control (C2) theory: a challenge to control science, IEEE Trans. Aut. Control, AC-32: 286.

Basar, T., and Haurie, A. (1984). Feedback equilibria in differential games with structural and modal uncertainties, in Advances in Large Scale Systems, J.B. Cruz Jr. Ed., JAI Press Corp., New York, p.163.

Basar, T., and Olsder, G.J. (1982). Dynamic Non-Cooperative Game Theory, Academic Press, New York.

Bensoussan, A., and Lions, J.L. (1975). Nouvelles méthodes en contrôle impulsionnel, J. Appl. Math. Optim., 1; 298.

Brémaud, P. (1981). Point Processes and Queues Martingale Dynamics, Springer-Verlag, New York.

Castanon, D.A., Coderch, M., Levy, B.C., and Willsky, A.S. (1980). Asymptotic analysis, approximation and aggregation methods for stochastic hybrid systems, Proc. American Control Conf., San Francisco, paper TA3-D.

Chizeck, H.J., and Lalonde, R. (1984). Discrete-time jump linear quadratic controlers with semi-markov transitions, Proc. 20th Asilomar Conf. Signals Comp. Systems, Pacific Grove, pp.61-65.

Ezzine, J., and Johnson, C.D. (1986). Analysis of continuous/discrete model parameter sensitivity via a perturbation technique, Proc. 18th Southeastern Symp. System Theory, pp.545-550.

Ezzine, J., and Haddad, A.H. (1988). On the stabilization of two-form hybrid systems via averaging, Proc. 22nd Conf. Inf. Sciences and Systems, Princeton.

Feller, W. (1966). An Introduction to Probability Theory and its Applications, vol. 2, J. Wiley, New York.

Isaac, R. (1965). Differential Games, J. Wiley, New York.

Malhame, R. (1988). On the role of renewal theory in the analysis of a class of cyclic hybrid state stochastic systems, Proc. 27th IEEE Conf. Decision Control, Austin.

Mariton, M. (1988). On systems with non markovian regime changes, to appear IEEE Trans. Aut. Control.

Mookerjee, P., Bar-Shalom, Y., and Campo, L. (1987). Estimation in systems with semi-markov switching models, Proc. 26th IEEE Conf. Decision Control, Los Angeles, pp.332-334.

Mortensen, R.E., and Haggerty, K.P. (1987). A stochastic computer model for heating and cooling loads, AIAA Power Engineering Society Summer Meeting, San Francisco, also to appear AIAA Trans. on Power Engineering.

Robinson, V.G., and Sworder, D.D. (1974). A computational algorithm for design of regulators for linear jump parameter systems, IEEE Trans. Aut. Control, AC-19: 47.

Rudemo, M. (1973). State estimation for partially observed markov chains, J. Math. Anal. Appl., 44: 581.

Sworder, D.D., and Robinson, V.G. (1973). Feedback regulators for jump parameters systems with state and control dependent transition rates, IEEE Trans. Aut. Control, AC-18: 355.

Sworder, D.D. (1980). Control of a linear system with non markovian modal changes, J. Econ. Dyn. Control, 2: 233.

Sworder, D.D. (1982). Regulation of stochastic systems with wide-band transfer functions, IEEE Trans. Syst. Man Cyber., SMC-12: 307.

Sworder, D.D., and Elliott, D.F. (1986). A renewal process model of nuclear background, Proc. 20th Asilomar Conf. Signals Comp. Systems, Pacific Grove.

Von Neuman, J., and Morgenstern, O. (1944). Theory of Games and Economic Behavior, Princeton University Press, Princeton.

Von Stackelberg, H. (1934). The Theory of Market Economy, trad. 1952, Oxford University Press, Oxford.

Wong, E., and Hajek, B. (1985). Stochastic Processes in Engineering Systems, Springer Verlag, New York.

Appendix 1
Stochastic Processes and Dynamic Systems

This appendix recalls the results in probability and stochastic processes theory used in the main body of the book. We shall follow quite closely (Wong and Hajek, 1985) and the reader will find more detailed treatments in (Feller, 1970) on probability, (Doob, 1953) on stochastic processes and (Friedman, 1975) on stochastic dynamic systems. The textbook (Arnold, 1974) is also used on stochastic differential equations and additional specialized references are indicated as we proceed.

A.1.1 Basic Probability

Probability theory is a mathematical construction to deal with physical systems driven by chance. The simplest situation is a *random experiment* (or trial) with a finite number of outcomes. We note $\Omega = \{\omega_1, \omega_2, ..., \omega_N\}$ the set of possible outcomes (or *sample space*) and $\omega \in \Omega$ a generic element of Ω. When casting a dice Ω has 6 elements for the six faces of the dice. To each ω_i we assign a probability p_i, imposing that $p_i \geq 0$, i = 1 to N, and $\sum_{i=1}^{N} p_i = 1$. To every subset S of Ω is then associated a *probability* defined as the sum of the probabilities of the outcomes contained in S. From the family of all subsets of Ω we extract a subfamily Ξ, called the *family of events* , corresponding to the subsets of Ω that are of interest to us, for example events {1, 3, 5} and {2, 4, 6} if we play an odd/even dice game. There are a few simple properties that Ξ should have to arrive at a satisfactory theory of probability. For example, the complement of an event or the intersection of two events should also be an event (non occurrence, simultaneous occurrence). Thus Ξ should be closed under all Boolean set operations (complementation, union, intersection; in fact it is enough to consider complementation and union or intersection). Since we want to consider sequences of events and possible convergence of such sequences, closure must extend to all countable operations. A family of sets closed under all countable Boolean set operations is called a *σ-algebra.*

A *probability measure* $\mathscr{P}$ is a function defined on a σ-algebra Ξ such that $0 \leq \mathscr{P}\{E\} \leq 1$ for any event E, $\mathscr{P}\{\Omega\} = 1$ and the following condition is satisfied:

if $\{E_j\}$ is a sequence of pairwise disjoint events in Ξ with $\bigcup_{j=1}^{\infty} E_j$ also

in Ξ then $\mathscr{P}\{\bigcup_{j=1}^{\infty} E_j\} = \sum_{j=1}^{\infty} \mathscr{P}\{E_j\}$

The fundamental concept is thus that of a *probability space*, i.e. a triplet $(\Omega, \Xi, \mathscr{P})$ with Ω the sample space, Ξ the σ-algebra of events of interest and $\mathscr{P}$ a probability measure on Ξ. A null set is associated with zero probability and the certain event, Ω, has probability one. The probability space is *complete* if every null set is an event. If this is not the case it can always be uniquely completed by extending $\mathscr{P}$.

The mathematical framework for the construction of probability spaces is measure theory. The pair (Ω, Ξ) is called a *measurable space* and a *measure* on Ξ is the generalization of the above probability measure when we drop the requirement that $\mathscr{P}\{\Omega\} = 1$. Consider two measurable spaces (Ω_1, Ξ_1) and (Ω_2, Ξ_2) and a function f from Ω_1 into Ω_2. The function f is called a *measurable mapping* if the inverse image of every event in Ω_2 is an event in Ω_1. A Ξ- measurable mapping X from (Ω, Ξ) into **R** with its Borel σ-algebra is called a (real) *random variable*. The *probability distribution function* of X is the function P_x defined by

$$P_x(a) = \mathscr{P}\{\{\omega, X(\omega) < a\}\}$$

We call the *expectation* of X the Stieltjes integral

$$\int_{-\infty}^{+\infty} x dP_x(x)$$

and denote it by E{X}. To stress that E{X} is indeed the integral of X over Ω with respect to the probability measure $\mathscr{P}$ we could use the notation

$$E\{X\} = \int_{\Omega} X(\omega) \mathscr{P}\{d\omega\}$$

(see (Breiman, 1968) for a precise definition of the latter integral). The *probability density function* of X, when it exists, is the non negative function $p_x(x)$ such that $dP_x(x) = p_x(x)dx$.

Two events E_1 and E_2 are *independent* if $\mathcal{P}\{E_1 \cap E_2\} = \mathcal{P}\{E_1\}\mathcal{P}\{E_2\}$. When $\mathcal{P}\{E_2\} > 0$ we define the *conditional probability* of E_1 given E_2 as

$$\mathcal{P}\{E_1 \mid E_2\} = \frac{\mathcal{P}\{E_1 \cap E_2\}}{\mathcal{P}\{E_2\}}$$

Obviously this reduces to $\mathcal{P}\{E_1\}$ if and only if the two events are independent. An *atom* A of a σ-algebra Ξ is a set of Ξ such that no subset of A, except A and the empty set, belongs to Ξ. Atoms are the irreducible components of a σ-algebra. Let X be a random variable with a well-defined expectation. For a sub-σ-algebra $\Xi' \subset \Xi$ the *conditional expectation* of X with respect to Ξ', noted $E\{X \mid \Xi'\}$ is the Ξ'-measurable random variable such that, for all A in Ξ',

$$E\{I_A\, E\{X \mid \Xi'\}\} = E\{I_A\, X\}$$

where I_A is the indicator function of A (= 1 if $X(\omega) \in A$, = 0 otherwise). Intuitively we "average" X on the atoms of Ξ'. In fact $E\{X \mid \Xi'\}$ is defined uniquely only up to null sets of Ξ'. The existence of this random variable results from the Radon-Nykodim theorem.

Consider a measurable space (Ω, Ξ) and two measures μ and μ_0. We say that μ is *absolutely continuous* with respect to μ_0 if every set of zero μ_0-measure also has a zero μ-measure. Then there exists a Ξ-measurable function Λ such that for every A in Ξ

$$\mu(A) = \int_A \Lambda(\omega)\, \mu_0(d\omega)$$

We call Λ the *Radon-Nikodym derivative* of μ with respect to μ_0 and write $\Lambda = d\mu/d\mu_0$. Again Λ is unique up to sets of zero μ_0-measure. The smoothing properties of conditional expectation include $E\{YX \mid \Xi'\} = YE\{X \mid \Xi'\}$ if Y is Ξ'-measurable and $E\{E\{X \mid \Xi_1\} \mid \Xi_2\} = E\{X \mid \Xi_2\}$ if $\Xi_2 \subset \Xi_1$ (both equalities hold almost surely, i.e. except maybe on $\mathcal{P}$- null sets).

Finally we introduce some convergence concepts that are later related to stability definitions (appendix 2). The convergence of a sequence $\{X_n\}$ of random variables can be defined in several ways. We shall say that it *converges almost surely* to X if there exists a (null) set S such that $\mathcal{P}\{S\} = 0$ and

$$\lim_{n\to\infty} | X_n(\omega) - X(\omega) | = 0 \quad \text{for every } \omega \notin S$$

We have *convergence in probability* if for every $\varepsilon > 0$ $\sup_{m\geq n} \mathcal{P}\{| X_m - X_n | \geq \varepsilon\} \underset{n\to\infty}{\to} 0$ and *pth mean convergence* if $\lim_{n\to\infty} E\{| X_n - X |^p\} = 0$.

A.1.2 Random Processes

By a *random* or *stochastic process* we mean a family of random variables $\{X_t, t \in T\}$, indexed by a real parameter t and defined on a common probability space $(\Omega, \Xi, \mathcal{P})$. The parameter t usually represents time and T is the time interval of interest $[t_0, t_f]$. For each $t \in T$, X_t is a Ξ-measurable function and by $\{X_t(\omega), t \in T\}$ we designate the *sample function* of the stochastic process associated with the event $\omega \in \Xi$. We shall not always use different notations for the process $\{X_t, t \in T\}$ and its value at time t X_t.

A process is *continuous in probability* at t if

$$\lim_{h\to 0} \mathcal{P}\{ |X_{t+h} - X_t| > \varepsilon\} = 0$$

for every $\varepsilon > 0$. It is *continuous in the pth mean* at t if

$$\lim_{h\to 0} E\{|X_{t+h} - X_t|^p\} = 0$$

It is *almost surely continuous* at t if

$$\mathcal{P}\{\omega: \lim_{h\to 0} |X_{t+h}(\omega) - X_t(\omega)| = 0\}\} = 1$$

If $X_t(\omega) = \lim_{\substack{s\to t\\ s>t}} X_s(\omega)$ and if $X_{t^-}(\omega) = \lim_{\substack{s\to t\\ s<t}} X_s(\omega)$ with probability 1 for all t then the

process is said to have continuous-on-the-right with limit-on-the-left sample paths, or simply that it is a *corlol process*. On the real line the Lebesgue measure is defined by extending the length measure of intervals, and we denote by **L** the σ-algebra of Lebesgue measurable sets in T. A stochastic process $\{X_t, t \in T\}$ is *separable* if there exists a countable set $S \subset T$ such that for any closed set $K \subset [-\infty, +\infty]$ and any open interval I, the sets

$$\{\omega: X_t(\omega) \in K, t \in I \cap T\} \text{ and } \{\omega: X_t(\omega) \in K, t \in I \cap S\}$$

differ by a fixed subset of zero $\mathcal{P}$- measure. The process $\{X_t, t \in T\}$ with a Lebesgue measurable time set T is *measurable* if $X_t(\omega)$ as a function of (t, ω) is measurable with respect to $\mathbf{L} \otimes \Xi$. Any process that is continuous in probability can be replaced, up to $\mathcal{P}$ null sets, by a measurable (and separable) counterpart. The process is *adapted* to an increasing family Ξ of sub-σ-algebras ($\Xi_s \subset \Xi_t$ for $s < t$) if X_t is Ξ_t-measurable for each t. Clearly X is adapted to the σ-algebra generated by the family of random variables $\{X_s, s \leq t\}$, often called the (internal) *history* of X. A non negative random variable τ is called a Ξ-*stopping time* if $\phi\{\tau \leq t\}$, with φ the indicator function, is adapted to Ξ. Let Π_t be the σ-algebra over $(0, \infty) \times \Omega$ generated by the rectangles $(s, t] \times R$, $0 \leq s \leq t$, $R \in \Xi_s$. A process X_t such that X_0 is Ξ_0- measurable and such that the mapping $(t, \omega) \to X_t(\omega)$ is Π_t- measurable, is said to be *predictable*.

Gaussian processes provide mathematical models for many random physical phenomena and a specific gaussian process, the brownian motion, is the basis for a rigorous understanding of the "white noise" notion. A random variable X is (non degenerate) gaussian if $E\{X^2\} < \infty$ and if it has a density function

$$p_x(x) = \frac{1}{\sqrt{2\pi\sigma^2}} \exp\left(-\frac{1}{2}\frac{(x-m)^2}{\sigma^2}\right)$$

where $m = E\{X\}$ and $\sigma^2 = E\{(X-m)^2\}$. A stochastic process $\{X_t, t \in T\}$ is a *gaussian process* if every finite linear combination of the form $Y = \sum_{i=1}^{N} \xi_i X_{t_i}$ is a gaussian random variable. The function $m_t = E\{X_t\}$ is called the mean function of the process and $M_{ts} = E\{(X_t - m_t)(X_s - m_s)\}$ is called the covariance function. The *brownian motion*, or Wiener process as it is sometimes called, is a gaussian process $\{X_t, t \in T\}$ with $E\{X_t\} = 0$ and $E\{X_t X_s\} = \min\{t, s\}$. Brownian motion is an example of a process with independent increments (X_t and $X_{t+h} - X_t$ for $h > 0$ are independent variables). Another example is the

Poisson process. It is defined as a counting process $\{N_t, t \in T\}$, i.e. an increasing right-continuous integer-valued random process such that $\Delta N_t = N_t - N_{t^-}$ is either zero or one for all t and ω. It satisfies the independent increments condition for $N_t - N_s$ a Poisson random variable with mean t-s whenever t > s. Note the similarity between brownian and Poisson processes which is further illustrated in §A.1.3 with the optimal filter representation.

A process $\{X_t, t \in T\}$ is a *Markov process* if for increasing instants $t_1 < t_2 < \dots < t_n$ in T we have

$$\mathcal{P}\{X_{t_n} < x_n | X_{t_k} = x_k, k = 1, n-1\} = \mathcal{P}\{X_{t_n} < x_n | X_{t_{n-1}} = x_{n-1}\}$$

Brownian motion has the Markov property. The *transition function* $P_x(x, t | \xi, s)$, t > s, of a Markov process X satisfies the *Chapman-Kolmogorov equation.*

$$P_x(x, t | x_0, t_0) = \int_{-\infty}^{+\infty} P_x(x, t | \xi, s)\, dP_x(\xi, s | x_0, t_0) \qquad t_0 < s < t$$

For a Markov process the future, given the past and the present, is equal to the future given only the present. It is sometimes said that the present summarizes the past. More precisely the σ-algebra generated by future values of X, $\sigma\text{-}\{X_s, s \geq t\}$, and the σ-algebra generated by past values of X, $\sigma\text{-}\{X_s, s \leq t\}$, are independent given X_t. The transition function $P_x(x, t | \xi, s)$ is *stationary* if it depends only on t-s and not on t and s separately. Assuming that the state space S of X is an interval we denote by $\mathcal{R}$ the smallest σ-algebra containing all subintervals of S (Borel sets). For g a complex-valued bounded function on S, measurable with respect to $\mathcal{R}$, we introduce an operator H_t by

$$(H_t g)(\xi) = E\{g(X_t) | X_0 = \xi\}$$

From the Chapman-Kolmogorov equation we deduce the *semigroup property* of the family of operators $\{H_t, 0 \leq t \leq \infty\}$

$$H_{t+s} = H_s H_t = H_t H_s$$

Provided the limit exists, the (infinitesimal) *generator* $\mathcal{L}$ of the Markov process X is defined as the linear operator

$$\mathscr{L} g = \lim_{t \to 0} (1/t)\,(H_t g - g)$$

where the limit is in the sense of strong convergence (Dynkin, 1965). For a Brownian motion the generator is computed as $\mathscr{L} g(\xi) = \frac{1}{2}\, g''(\xi)$ where g" is the bounded-continuous second derivative of g.

To make the concept of differential equations driven by white noise precise we need to introduce stochastic integrals, that is, a quantity of the form

$$\int_{t_0}^{t_f} g(\omega, t)\, db(\omega, t)$$

where b is a brownian motion. Since b has unbounded variations this integral cannot be defined in the usual sense. We shall introduce the generally accepted definition due to Ito and the associated differentiation rules. The importance of the Ito integral is that it produces martingales, a useful property as we shall see in §A.1.3 when studying optimal filters. We first impose some restrictions on the integrand g: it is assumed to be jointly measurable in (ω, t) with $\int_{t_0}^{t_f} E\{|\, g\, |^2\}\, dt < \infty$. For instants $t_0 < t_1 < \dots < t_n = t_f$ independent of ω with $\Delta = \max_i t_{i+1} - t_i$ we define the *Ito stochastic integral* by

$$\int_{t_0}^{t_f} g(\omega, t)\, db_t(\omega) = \lim_{\Delta \to 0} \sum_{k=1}^{n-1} g(\omega, t_k)\,(b_{t_{k+1}}(\omega) - b_{t_k}(\omega))$$

Despite the use of the same notation ($\int$) the stochastic integral does not behave like its ordinary counterpart, for example $\int_0^t b_s db_s$ is not equal to $\frac{1}{2}\, b_t^2$. Using the Ito integral we can interpret the *stochastic differential equation*

$$dX_t(\omega) = f(X_t(\omega), t)\, dt + \sigma(X_t(\omega), t)\, db_t(\omega)$$

with initial condition $X_{t_0} = X_0$. We say that this generates the process $X_t(\omega)$ as

$$X_t = X_0 + \int_{t_0}^{t} f(X_s, s)\, ds + \int_{t_0}^{t} \sigma(X_s, s)\, db_s$$

for suitable smoothness restrictions on f and σ. The most important properties of the process thus defined are that it is sample continuous, unique with probability 1 and that it has the Markov property.

For X_t generated by $dX_t = f_t dt + \sigma_t db_t$ and $Y_t = F(X_t, t)$ some transformation of it, the *Ito differentiation rule* gives us the *correction term* to write Y_t as an integral

$$Y_t = Y_0 + \int_{t_0}^{t} \dot{F}(X_s, s)\, ds + \int_{t_0}^{t} F_x(X_s, s)\, dX_s + \frac{1}{2} \int_{t_0}^{t} F_{xx}(X_s, s)\ \sigma_s^2\, ds$$

where $\dot{F} = \partial F/\partial t$, $F_x = \partial F/\partial X$ and $F_{xx} = \partial^2 F/\partial X^2$. The last (correction) term gives the corrected value of $\int_0^t b_s db_s$. For $Y_t = \frac{1}{2} b_t^2$, we obtain $\frac{1}{2} b_t^2 = \int_0^t b_s db_s + \frac{1}{2} t$.

The next kind of random process we shall introduce is the martingale process. It plays an important role both in optimal filtering (see §A.1.3) and in stochastic optimal control (Fleming and Rishel, 1975, Hijab, 1986). A *martingale* with respect to $\{\Xi_t, t \in T\}$ is a stochastic process X adapted to $\{\Xi_t, t \in T\}$ and satisfying

$$E\{X_t \mid \Xi_s\} = X_s \quad \text{almost surely for } t \geq s$$

The process is said to be a *supermartingale* (resp. *submartingale*) if equality is replaced by $\leq$ (resp. $\geq$).

If X is a separable martingale and $t_f < \infty$ then we have Doob's L^2 inequality

$$E\{\sup_{0 \leq t \leq t_f} |X_t|^2\} \leq 4E\{|X_{t_f}|^2\}$$

Also if $E\{X_t^2\} < \infty$ for every $t \in [0, t_f]$ then for every positive ε,

$$\mathcal{P}\{\sup_{0 \leq t \leq t_f} |X_t| \geq \varepsilon\} \leq E\{| X_{t_f}^2\}/\varepsilon^2$$

In fact there is a wealth of other inequalities involving L^p norms and super/sub-martingales (see the monograph (Garsia, 1973)). In particular a useful inequality is obtained for X_t a supermartingale and λ a positive number

$$\mathcal{P}\{\sup_{0 \leq t \leq t_f} |X_t| \geq \lambda\} \leq |X_0| / \lambda$$

For example, if w is a brownian motion adapted to an σ-algebra Ξ and if $w_s - w_t$, $s \geq t$ is independent of Ξ_t for any t, then it is a martingale with respect to Ξ, often called a Wiener Ξ-martingale. Also, if N is a Poisson process adapted to Ξ with $N_s - N_t$, $s \geq t$ independent of Ξ_t for any t then $n_t = N_t - t$ is a Ξ-martingale, often called a Poisson martingale. The definition of stochastic integrals given above for brownian motion can be generalized to a class of martingales and the result of integration is again a martingale.

A random process X is called a *local martingale* if it is corlol and if there is a sequence τ_n of increasing stopping times ($\tau_n(\omega) \leq \tau_{n+1}(\omega)$ w.p.1) with $\lim_{n \to \infty} \tau_n = +\infty$ almost surely such that X^{τ_n} is a martingale for each n, where $X_t^{\tau_n} = 0$ if $\tau_n = 0$, $X_t^{\tau_n} = X_t$ if $0 \leq t \leq \tau_n$ and $X_t^{\tau_n} = X_{\tau_n}$ if $t \geq \tau_n$. Building on Girsanov's theorem (Van Shuppen and Wong, 1974), a fundamental *martingale representation* result can be obtained (Lipster and Shiryayev, 1976). For w a Wiener process and N a Poisson process, let F_t be the increasing family σ-$\{w_s, N_s, s \leq t\}$ completed by the null sets of $(\Omega, \Xi, \mathcal{P})$. Then any corlol local F_t martingale has the representation

$$X_t = X_0 + \int_0^t \Gamma_s^w \, dw_s + \int_0^t \Gamma_s^n \, dn_s$$

where $n_t = N_t - t$ and Γ^w and Γ^n are predictable.

A.1.3 Optimal Filtering

Filtering is the name given to bayesian estimation of the state at time t of a dynamical stochastic system, based on the incomplete information contained in measurements, or observations, available at time t. If data are available up to $t - \tau$ (resp. $t + \tau$) for some positive τ we speak of a *prediction* (resp. *smoothing*) problem. These problems were first studied in terms of frequency spectra leading to the "Wiener filter"

(Wiener, 1949) and Kalman showed that a time domain approach leads to a solution (the "Kalman filter") that is easier to compute and implement (Kalman, 1960).

The main advantage of the time domain solution is that it is expressed in *recursive* form: denoting by Ξ_t the underlying (global) σ-algebra on $(\Omega, \Xi, \mathscr{P})$ we have an observation σ-algebra O ($O_t \subset \Xi_t$) generated by the observations o_t as σ-$\{o_t, t_0 \le s \le t\}$. Usually Ξ_t has a representation $O_t \vee X_t$ where X_t is the σ-algebra generated by the state of the system x_t as σ-$\{x_s, t_0 \le s \le t\}$. For $f(x_t)$ some integrable process adapted to Ξ_t we are interested in finding a way of computing the estimate $\hat{f}(x, t) = E\{f(x_t) \mid O_t\}$. Loosely speaking, what recursive means is that we would like to have $\hat{f}(x, t + dt) = g(\hat{f}(x, t), do_t, t)$. In other words, the estimate at time t + dt should depend only on the estimate at time t and on the new information contained in the observations increment $do_t = o_{t+dt} - o_t$. Recursiveness saves computer memory because the current estimate summarizes the past and the estimation algorithm $\hat{f}$ combines it with do_t to produce a new estimate. Note that we have right away restricted ourselves to estimates of the form $E\{- \mid O_t\}$, that is to expectations conditioned on the observations. As is well known, this provides an optimal estimate in the sense of *mean square error* $E\{ \mid f - \hat{f}\mid^2\}$, hence the expression *optimal filter*.

Martingale theory provides a unifying framework to study optimal filtering for Wiener-driven systems and point-process systems (Brémaud, 1981, Elliott, 1982). This is important for applications where we often find the two types of processes intermingled, in particular with the hybrid models studied in this book.

We now recall the main steps leading to the optimal filter representation. Consider the process

$$X_t = X_0 + \int_{t_0}^{t} f_s \, ds + \mu_t$$

with $E\{|X_0|\}$ finite, f a Ξ_t-progressive process, $E\{ \int_{t_0}^{t} |f_s| ds\} < \infty$ and μ a Ξ-martingale.

Then, for O the observation σ-algebra, there exists an O-martingale ν such that for each t

$$\hat{X}_t = \hat{X}_0 + \int_{t_0}^{t} \hat{f}_s \, ds + \nu_t \quad \text{almost surely}$$

With reference to the Hilbert space theory of linear filters for second order processes (Wiener, 1949) this is often called the *representation of projections* onto O. We have Wiener-driven (y) and point process (z) observations

$$y_t = \int_{t_0}^{t} h_s \, ds + w_t, \qquad \text{w Wiener martingale}$$

(with respect to $O_t \vee \Xi_{tf}$)

$$z_t = \int_{t_0}^{t} \lambda_s \, ds + M_t, \qquad \text{M Poisson martingale}$$

for h progressive and λ predictable with respect to Ξ. We impose almost sure conditions on h and λ

$$E\{ \int_{t_0}^{t} h_s^2 \, ds\} < +\infty$$

$$E\{ \int_{t_0}^{t} \lambda_s \, ds\} < \infty, \ \lambda_s > 0$$

For $O_t = Y_t \vee Z_t$ (completed) with $Y_t = \sigma\text{-}\{y_s, t_0 \le s \le t\}$ and $Z_t = \sigma\text{-}\{z_s, t_0 \le s \le t\}$ we would like to compute the optimal estimate of the state ξ_t of the system

$$\xi_t = \xi_0 + \int_{t_0}^{t} f_s \, ds + \mu_t, \qquad \mu \ \Xi\text{-martingale}$$

with f predictable with respect to Ξ, ξ_0 Ξ_0-measurable, $E\{|\xi_0|\} < \infty$ and

$$E\{ \int_{t_0}^{t} |f_s| \, ds\} < +\infty$$

The O-martingales

$$\tilde{y}_t = y_t - \int_{t_0}^{t} \hat{h}_s \, ds$$

$$\tilde{z}_t = y_t - \int_{t_0}^{t} \hat{\lambda}_s \, ds$$

are called the *innovation processes*. Using the martingale representation theorem, the optimal estimate is then obtained as

$$\hat{\xi}_t = \hat{\xi}_0 + \int_{t_0}^{t} \hat{f}_s \, ds + \int_{t_0}^{t} \Gamma_s^w \, d\tilde{y}_s + \int_{t_0}^{t} \Gamma_s^n \, d\tilde{z}_s$$

where Γ_s^w and Γ_s^n are the predictable processes given, almost everywhere, by

$$\Gamma_s^w = E\{(\xi_s - \hat{\xi}_s)(h_s - \hat{h}_s) \mid O_t\} + E\{\psi_s^w \mid O_t\}$$

$$\Gamma_s^n = \frac{E\{(\xi_s - \hat{\xi}_s)(\lambda_s - \hat{\lambda}_s) \mid O_{t-}\} + E\{\psi_s^n \mid O_{t-}\}}{E\{\lambda_s \mid O_{t-}\}}$$

for ψ^w and ψ^n Ξ predictable processes such that

$$\int_{t_0}^{t} \psi_s^w \, ds = \langle \mu, w \rangle_t$$

$$\int_{t_0}^{t} \psi_s^n \, ds = \langle \mu, M \rangle_t$$

The integrands Γ_s^w and Γ_s^n are called the *innovation gains*.

A.1.4 Markov Chains

We call a homogeneous *Markov chain* the Markov process X_t with values in $\mathbf{N}^+$ and transitions $\mathscr{P}\{X_{t+s} = j \mid X_s = i\} = q_{ij}(t)$ for a given infinite matrix $Q_t = (q_{ij})_{i,j \in \mathbf{N}^+}$.

Under the continuity condition $\lim_{t \to 0} q_{ij}(t) = \delta_{ij}$ where δ is the Kronecker index ($\delta_{ij} = 1$ if i = j, 0 otherwise), it can be shown (Freedman, 1983) that

$$\lim_{t \to 0} \frac{1 - q_{ii}(t)}{t} = \pi_i \leq \infty$$

$$\lim_{t \to 0} q_{ij}(t)/t = \pi_{ij} < \infty$$

If for every $i \in \mathbf{N}^+$ π_i is finite the chain is *stable*. If $\pi_i = \sum_{j \neq i} \pi_{ij}$ it is *conservative*. The set of parameters $\Pi = (\pi_{ij})_{i,j \in \mathbf{N}^+}$ with, by convention, $\pi_{ii} = -\pi_i$ gives the infinitesimal generator of the Markov process X_t and the π_{ij}s are called the *infinitesimal characteristics*. For a conservative homogeneous Markov chain the (forward) Chapman-Kolmogorov equation is

$$\dot{q}_{ij} = - q_{ij} \pi_j + \sum_{k \neq j} q_{ik} \pi_{kj}, \quad i, j \in \mathbf{N}^+$$

The Markov chain is *finite* when X_t takes only a finite number of values $X_t \in \mathbf{S}$ with $\mathbf{S} = \{1, 2, ..., N\}$.

The chain can also be described in terms of its indicator ϕ_t, a vector in $\mathbf{R}^N$, with $\phi_{ti} = 1$ if $X_t = i$, 0 otherwise, i = 1 to N. The process M_t defined by $dM_t = d\phi_t - \Pi' dt$ is a martingale with respect to σ-$\{X_s, s \leq t\}$ (Brémaud, 1981).

If Π does not depend on time the chain is *stationary*. We define the time translation operator T_t by $T_t X_s = X_{t+s}$ and Y is an *invariant random variable* of X if, for each t, Y and $T_t Y$ differ only on a set of zero $\mathcal{P}$- measure. The chain is *ergodic* if all its invariant random variables are almost surely constant. For an ergodic process we know that the ensemble average equals the time average.

The elements of **S** are often called the states of the Markov chain but in the context of hybrid systems we more often speak of regimes. A regime i is *recurrent* if for any instant s $\mathcal{P}\{X_t = i \mid X_s = i\} = 1$ for some t > s. Otherwise it is said to be *transient*. If the return time $T_R^i = \text{Inf}\{t > 0, X_t = i, X_0 = i\}$ has finite expectation for all i = 1 to N then the chain is said to be *positive recurrent*. A regime i with $\pi_{ii} = 0$ is called *absorbing*. A regime i is *accessible* from regime j if it is possible to begin in j and arrive in i within some finite time. A *communicating class* is a set of regimes all accessible from eachother. A communicating class is *closed* if exit is not possible. An *irreducible* chain is such that for each pair (i, j) in S^2 and any instant s there exists an instant t > s such that $\mathcal{P}\{X_t = j \mid X_s = i\} > 0$. A chain with a fixed sequence of regimes (but random transition times) is called a *directed chain*. In addition to (Freedman, 1983), a basic reference on finite Markov chains is (Kemeny and Snell, 1960).

References

Arnold, L. (1974). Stochastic Differential Equations. Theory and Applications, J. Wiley, New York.

Breiman, L. (1968). Probability, Addison-Wesley, New York.

Brémaud, P. (1981). Point Processes and Queues - Martingale Dynamics, Springer Verlag, New York.

Doob, J.L. (1953). Stochastic Processes, J. Wiley, New York.

Dynkin, E.B. (1965). Markov Processes, vol.1 and 2, Springer Verlag, Berlin.

Elliott, R.J. (1982). Stochastic Calculus and Applications, Springer Verlag, New York.

Feller, W. (1970). An Introduction to Probability Theory and its Applications, J. Wiley, New York.

Fleming, W.H., and Rishel, R.W. (1975). Deterministic and Stochastic Optimal Control, Springer Verlag, New York.

Freedman, D. (1983). Markov Chains, Springer Verlag, New York.

Friedman, A. (1975). Stochastic Differential Equations and Applications, vol.1 and 2, Academic Press, New York.

Garsia, A.M. (1973). Martingale Inequalities, W.A. Benjamin, Reading.

Hijab, O. (1986). Stabilization of Control Systems, Springer Verlag, New York.

Kalman, R.E. (1960). A new approach to linear filtering and prediction problems, Trans. ASME, J. Basic Eng., 82: 35.

Kemeny, J.G., and Snell, J.L. (1960). Finite Markov Chains, Van Nostrand, Princeton.

Lipster, R., and Shiryayev, A. (1978). Statistics of Random Processes, vol. 1 and 2, Springer Verlag, Heidelberg.

Van Schuppen, J.H., and Wong, E. (1974). Transformation of local martingales under a change of law, Annals of Prob., 2: 878.

Wong, E., and Hajek, B. (1985). Stochastic Processes in Engineering Systems, Springer Verlag, New York.

Wiener, N. (1949). Time Series, MIT Press, Cambridge.

Appendix 2
Stochastic Control: Stability and Optimization

Control problems for stochastic systems were studied extensively over the past twenty years. In this appendix we summarize the main results needed in the rest of the book, focusing on stability and optimal control. Our main reference on stochastic stability is (Kushner, 1967) and we also borrow from (Arnold and Wihstutz, 1986) on the concept of Liapunov exponents. Related monographs are (Aström, 1970, Has'minskii, 1980) and we shall quote other original references as we proceed. The presentation of dynamic programming conditions for optimally controlled markov processes is adapted from (Fleming and Rishel, 1975). We briefly discuss the extension of Pontryagin's maximum principle to stochastic systems (Pontryagin et al. , 1962). It was found useful in (Sworder 1968, 1969) when designing jump linear quadratic regulators. Finally we mention a tool that permit an easy derivation of optimality conditions in constrained design, the matrix maximum principle (Athans, 1968) and the corresponding list of gradients (Berger, 1976, Brewer, 1977).

A.2.1 Stochastic Stability

Consider first a deterministic system $\dot{x}_t = f(x_t, t)$ with initial condition x_0 belonging to a region R_0 of the state space. Stability of the system at the origin $x = 0$, assumed to be an equilibrium point, can be understood as the convergence of x_t to 0 as time goes, $\lim_{t\to\infty} x_t = 0$. Obviously for non linear dynamics it is necessary to discuss this limit for each initial condition and practical considerations also suggest that we would be interested in having some information on the rate at which the equilibrium is attained. To limit computations, it would be nice to get answers for this type of stability questions without actually integrating the model. The purpose of the qualitative theory of differential equations is precisely this and its two basic results that we shall use are Liapunov functions and Liapunov exponents. However we study primarily stochastic systems and we need first to recall stochastic stability definitions. Depending on the way randomness is taken into account there are several stability notions associated with a given deterministic notion. For simplicity we take the origin $x = 0$ as the equilibrium point of interest.

The origin is *stable with probability one* if for any $\varepsilon > 0$ and $\rho > 0$ there exists $\delta > 0$ such that $\| x_0 \| < \delta$ implies $\mathcal{P} \{ \sup_{0 \le t < \infty} \| x_t (x_0) \| \ge \varepsilon \} \le \rho$. It is *asymptotically stable with probability one* if in addition $\lim_{t \to \infty} x_t (x_0) = 0$ with probability one for all x_0 in some neighbourhood of the origin. If this neighbourhood covers the entire state space we add *complete*, or *in the large*. We say that the origin is *exponentially stable with probability one* if it is stable with probability one and if there exists a and b, positive and finite such that

$$\mathcal{P} \{ \sup_{0 \le \tau < \infty} \| x_\tau (x_0) \| \ge \varepsilon \} \le b e^{-at}$$

The process x_t is *uniformly bounded by* ε *with probability* ρ if

$$\mathcal{P} \{ \sup_{0 \le t < \infty} \| x_t (x_0) \| \ge \varepsilon \} \le 1 - \rho$$

and we say that it is uniformly bounded with probability one and with bound m if, for each x_0,

$$\lim_{t \to \infty} \mathcal{P} \{ \sup_{0 \le \tau < \infty} \| x_\tau (x_0) \| > m \} = 0$$

Sometimes it is more convenient to restrict attention to moments of the solution. We say that the origin is *stable in the pth mean* if, for any $\varepsilon > 0$ there exists $\delta > 0$ such that $\| x_0 \| < \delta$ implies $E \{ \| x_t(x_0) \|^p \} < \varepsilon$ for $t > 0$. It is *asymptotically stable in the pth mean* if in addition $\lim_{t \to \infty} E \{ \| x_t(x_0) \|^p \} = 0$. Finally if there further exists a and b positive such that

$$E \{ \| x_t(x_0) \|^p \} \le b \| x_0 \|^p e^{-at} \qquad \text{for } t > 0$$

then we have *exponential stability of the pth mean*.

The notion that comes closest to deterministic stability is almost sure stability.

We say that the origin is *almost surely stable* if $\mathcal{P}\{\lim_{\|x_0\|\to 0} \sup_{0\le t<\infty} \| x_t(x_0)\| = 0\} = 1$. It is *almost surely asymptotically stable* if in addition there exists a δ such that $\| x_0 \| < \delta$ implies, for any $\varepsilon > 0$, $\lim_{t\to\infty} \mathcal{P}\{\sup_{\tau\ge t} \| x_\tau(x_0)\| > \varepsilon\} = 0$.

The survey paper (Kozin, 1969) contains a discussion of the possible stability concepts for stochastic systems, complete with examples.

The stochastic Liapunov function plays the same role for the study of (controlled) Markov process as Liapunov functions do for deterministic stability analysis. It turns out that, roughly speaking, the key step is to prove that a candidate positive function of the system variables possesses the supermartingale property (see appendix 1). Bertram and Sarachik (1959) and Kats and Krasovskii (1960) originally proved moments stability by Liapunov function methods without reference to martingale theory.

We illustrate the use of stochastic Liapunov functions by quoting an exponential stability result, see (Kushner, 1967) for corresponding results regarding other types of stability. Consider a Markov process $\{x_t, t_0 \le t \le t_f\}$ with generator $\mathcal{L}$. We assume that the non negative and continuous candidate Liapunov function V(x) is such that V (0) = 0. Then if $\mathcal{L}V(x) \le -aV(x)$ for some $a > 0$ we have

$$\mathcal{P}\{\sup_{t\le\tau<\infty} V(x_\tau) \ge \lambda\} \le \frac{V(x)e^{-at}}{\lambda} \qquad \text{for } x_t = x$$

and the process is exponentially stable with rate at least a.

As in the deterministic case, a general difficulty is to find suitable Liapunov functions. Even for linear differential equations with Markov process coefficients it is hard to construct a family of Liapunov functions that would completely characterize stability of the solutions.

The Liapunov function can also be useful for designing control laws with some stabilizing property. This was indeed pursued for deterministic systems in (Kalman and Bertram, 1960). Essentially the control is chosen to ensure that the Liapunov function decreases in a neighbourhood of the equilibrium point and it generalizes to the stochastic case (see the examples of (Kushner, 1967)).

An alternative to Liapunov functions is to compute the Liapunov exponents (also called Liapunov characteristic numbers) of the solution. The basic idea is a comparison to the class of exponential functions. For a deterministic process x_t the

Liapunov exponent would be simply $\lambda = \lim_{t\to\infty} \frac{1}{t} \log \| x_t \|$ with, obviously $\lambda = a$ when $x_t = \exp at$. In the stochastic setting we write $x_t(x_0, \omega)$ to stress the dependence on both initial condition and chance.

The sample path Liapunov exponent $\lambda(x_0, \omega)$ is defined as

$$\lambda(\omega, x_0) = \lim_{t\to\infty} \frac{1}{t} \log \| x_t(x_0, \omega) \|$$

and we have the intuitive stability condition : the system is almost surely exponentially stable if and only if

$$\operatorname{Max}_{x_0 \neq 0} \lambda(\omega, x_0) < 0$$

Note that, contrary to Liapunov functions, Liapunov exponents lead to sufficient conditions that are also necessary. However it is in general difficult to compute $\lambda(\omega, x_0)$ analytically and we must resort to numerical approximations (Pardoux and Talay, 1985) . It is sometimes easier to evaluate exponents associated with the moments of the solution

$$\lambda_p (x_0) = \lim_{t\to\infty} \frac{1}{t} \log E \{\| x_t(x_0, \omega) \|^p\} \qquad p > 0$$

We have exponential stability of the pth moment when $\lambda_p < 0$. Using Liapunov exponents it has been possible to give a mathematically rigorous answer to the question of the influence of multiplicative noise on stability (e.g. Arnold et al. , 1983) : a damped linear oscillator with a spring constant driven by brownian motion can go from stable to unstable or from unstable to stable as the random term intensity is increased.

A.2.2 Stochastic Dynamic Programming

For a deterministic system we can deduce the state trajectory from the knowledge of the initial conditions and the control used up to the current time. Observing the current state therefore does not bring additional information. An important consequence is that closed-loop and open-loop control policies are equivalent for deterministic systems, and, in particular, an optimal feedback law does not give better performance than its open-loop counterpart.

This is no longer the case when the system is subject to imperfectly known disturbances and it is then important to precisely delineate the information structure available at the controller. We first consider the case of *complete observations* when all the plant variables are known at each instant of time and introduce the machinery of dynamic programming (Bellman, 1957, Fleming and Rishel, 1975). The controlled process is supposed to be Markov with generator $\mathcal{L}_u$, and we indicate the fact that the process is a controlled Markov process by adding u as an index on the generator $\mathcal{L}$ defined in appendix 1. The cost is defined as

$$J = E\{ \int_{t_0}^{t_f} L(t, x_t, u_t)\, dt \mid x_0, t_0\}$$

with L a continuous function satisfying the (polynomial) growth condition

$$| L(t, x_t, u_t) | \leq C\,(1 + | x_t | + | u_t |)^k$$

for suitable constants C and k. The value of J depends on the initial condition and on the control used over the optimization interval and we indicate this by writing $J = J\,(x_0, t_0, u)$.

We look for closed-loop control laws $u = U(t, x)$. The class $\mathcal{U}$ of admissible policies is defined by imposing smoothness conditions on U :

$$| U(t, x) | \leq M(1 + | x |)$$

$$| U(t, x) - U(t, y) | \leq K\, | x - y |$$

for positive constants M and K . The optimization problem is then to minimize J over the class of admissible polices.

Assuming the existence of an optimal solution, noted u^*, we consider the optimal expected performance as a function of the initial data x_0 and t_0 and we introduce the cost-to-go V(x, t) as the optimal cost for the problem with initial data x and t

$$V(x, t) = \inf_{u \in \mathcal{U}} J\,(x, t, u)$$

From the definition of the generator we have formally

$$V(x_0, t_0) = E\{V(x_t, t) - \int_{t_0}^{t} (V_t + \mathscr{L}_u V)\, d\tau \mid x_0, t_0\}$$

for $t_0 < t \leq t_f$ and $V_t + \mathscr{L}_u V$ evaluated at (x_τ, τ). Now assume that we use the optimal control $u^*(\tau, x)$ for $\tau > t$ and $u(\tau, x)$ for $\tau < t$. Then

$$V(x_0, t_0) \leq E\{ \int_{t_0}^{t} L(\tau, x_\tau, u_\tau)\, d\tau + V(x_t, t) \mid x_0, t_0\}$$

and equality holds if we use u^* over $[t_0, t_f]$. Letting t go to t_0 and comparing the two expressions of $V(x_0, t_0)$ we obtain the dynamic programming equation of optimal stochastic control

$$0 = V_t + \min_{u \in \mathscr{U}} (\mathscr{L}_u V + L)$$

with the boundary condition $V(x, t_f) = 0$ from the definition of the cost-to-go. Terminal cost could be included as well to get a non zero terminal condition. The above equation is familiarly referred to as Hamilton - Jacobi - Bellman equation. It is not straightforward to make the above derivation rigorous and this is usually done for restricted classes of processes. The case of controlled diffusions is detailed in (Fleming and Rishel, 1975) and optimal control of jump processes is considered in (Varaiya, 1975) or the monograph (Elliott, 1982). An interesting viewpoint is to observe that the optimal cost-to-go satisfies the martingale property

$$E\{V(x^*_\tau, \tau) \mid x^*_t)\} = V(x^*_t, t) \qquad \text{for } t_0 \leq t \leq \tau \leq t_f$$

where x^* denotes the optimal state trajectory under u^*. Using variational inequalities (Bensoussan and Lions, 1975) solve the problem by considering directly the operator $\mathscr{L}_u$.

Hamilton - Jacobi - Bellman equation is used as a verification theorem giving a sufficient condition for the optimal control : solve this second order differential equation in the cost-to-go and perform the indicated minimization to find u^*. Practically we first postulate the form of the cost-to-go to express u^*as a function of V_x and then try to satisfy the conditions imposed on V by the Hamilton - Jacobi - Bellman equation with its boundary data.

The above discussion assumed that the stochastic process x_t itself is known to the controller, in other words we assumed complete observations. However the more realistic situation is the case of partial observations when we have access to only part of the plant variables. A notion of deep practical significance is then the certainty equivalence principle. When it holds we can use our partial observations to generate estimates of unobserved variables and subsequently use these as if they were the exact (true) variables. Unfortunately we do not have certainty equivalence in general, with the exception of linear quadratic gaussian problems where it is well known (Anderson and Moore, 1971) that the Kalman filter estimates can be used in place of the plant state for the LQ regulation loop. Historically the certainty equivalence principle was coined in econometrics (Simon, 1956) and (Joseph and Tou, 1961) first mentionned the related separation theorem in an engineering context (see also (Wonham, 1968)). Using conditional expectations as estimators (see appendix 1), a general approach to the partially observed problem is to transform it into a completely observed problem in the estimated variables (Rishel, 1981). However the resulting problem is very difficult, see for example (Hijab, 1983) or (Caines and Chen , 1985). The work of (Witsenhausen, 1968) shows that controller memory can improve the performance at the cost of increased complexity.

A.2.3 On the Stochastic Maximum Principle

There has been many efforts to extend Pontryagin's maximum principle to the optimization of stochastic systems. From the russian school we can mention the monographs (Batkov, 1974, Kazakov, 1975) with an application to systems containing jump parameters in (Bukhalev, 1980). In fact the original work of (Pontryagin et al. , 1962) already studied some special stochastic problems.

Similar work has been carried out independently in (Kushner, 1965 a and b) , (Fleming, 1968) and a quite general formulation has been obtained in (Warfield, 1976). Sworder used his study of the stochastic maximum principle (Sworder, 1968) to derive the original JLQ regulator (Sworder, 1969).

For simplicity we consider a linear system with random parameters

$$\dot{x}_t = A(t, \omega)\, x_t + B(t, \omega)\, u_t$$

where ω displays explicitly the randomness of the A and B matrices. The cost function is

$$J = E\{ \int_{t_0}^{t_f} L(\tau, x_\tau, u_\tau)\, d\tau \mid x_0, t_0\}$$

with $x(t_0) = x_0$ the initial condition. The class of admissible control $\mathcal{U}$ is the set of all measurable functions such that the corresponding x_t is uniformly continuous and satisfies a local Lipchitz condition. In essence this simply says that any measurable control that produces a smooth state trajectory is admissible. Some restrictions are also imposed on the integrand L : it is continuous and positive and the gradients L_x and L_u must exist and be continuous.

A stochastic Hamiltonian is defined as

$$H = \lambda_t\, (x, \omega)\, (A(t, \omega)\, x + B(t, \omega)\, u) - L(t, x, u)$$

We then have the following stochastic analogue of Pontryagin's maximum principle :

If u^* denotes the optimal control and u is any other admissible control such that $E\,\{\lambda_t'\,(x, \omega)\, A(t, \omega)\, x_t + B(t, \omega)\, u_t\,\}$ is measurable then the event

$$E\{H(x, u, t, \omega) \mid x, t\} > E\,\{H(x, u^*, t, \omega) \mid x, t\}$$

has a zero probability (Sworder, 1968, Warfield, 1976). In other words the optimal control maximizes the conditional expectation of the Hamiltonian. The co-state vector λ satisfies a random differential equation, integrated backwards in time

$$\dot{\lambda}_t = - \partial H/\partial x\,, \qquad \lambda_{tf} = 0$$

For the above linear problem this gives

$$\dot{\lambda}_t = - (A(t, \omega) + B(t, \omega)\, u_t)'\, \lambda_t + (L_x + L_u\, u_x)'\,, \qquad \lambda_{tf} = 0$$

where u_x, L_x and L_u stand for partial derivatives with respect to the indicated variable. If the control is unconstrained the maximization of the Hamiltonian implies $\partial E\,\{H \mid x, t\} / \partial u = 0$. This can be solved for a quadratic cost ($L = x'\, Q_1\, x + u'\, Q_2\, u$) to obtain

$$u_t^* = \frac{1}{2} Q_2^{-1} E \{B' \lambda_t | x, t\}$$

The optimal control is thus formulated in terms of a stochastic generalization of Pontryagin's familiar two-point boundary-value problem

$$\dot{x}_t^* = A(t, \omega) x_t^* + \frac{1}{2} B(t, \omega) Q_2^{-1} E \{B(t, \omega)' \lambda_t^* | x_t, t\} \quad , x_{t0} = x_0$$

$$\dot{\lambda}_t^* = - (A(t, \omega) + \frac{1}{2} B(t, \omega) Q_2^{-1} E \{B(t, \omega)' \lambda_{xt}^* | x_t^*, t\})' \lambda_t^*$$

$$+ (2Q_1 x_t^* + 2E \{B(t, \omega)' \lambda_{xt}^* | x_t^*, t\})' \quad , \lambda_{tf} = 0$$

and, as in the deterministic case, a simple solution is available only for cases where a relation like $\lambda_t(x, \omega) = 2\Lambda_t(\omega)x$ exists (see chapter 3 or (Sworder, 1969, Bukhalev, 1980) for some examples). For a comparative discussion of maximum principle and dynamic programming conditions, see (Fleming and Rishel, 1975).

A.2.4 The Matrix Maximum Principle

In applications the engineer often has a good feeling, based on constraints, experience and intuition, on the structure of the control laws that should be used. Rather than optimizing the control law itself, it is then natural to restrict attention to the optimization of the free parameters of the desired structure, in other words to solve a constrained optimization problem. The most typical situation is when we want to find the gains of a linear feedback like $u_t = \Gamma y_t$ with y_t some measured variables. It is then interesting to have optimality conditions directly on Γ rather than on u and this is the purpose of the matrix maximum principle (Athans, 1968).

The system is given in matrix form as

$$\dot{X}_t = F(X_t, U_t)$$

where X (resp. U) is an nxn (resp. mxm) state (resp. control) matrix. We shall give an example below illustrating how this can be derived naturally from the more usual vector representation. Similarly the cost is an integral in terms of the matrices U and X

$$J = \int_{t_0}^{t_f} L(\tau, X_\tau, U_\tau)\, d\tau$$

The minimization of J subject to the dynamic constraint relating X and U is readily formulated in terms of the Hamiltonian

$$H = L(t, X_t, U_t) + \text{trace}\, \dot{X}_t P_t'$$

where P is an nxn co-state matrix. Optimality conditions are given through gradient matrices, defined as follows : for M a matrix and for f(M) a scalar function of the entries m_{ij} of M, supposed to be independent, the gradient of f with respect to M, denoted $\partial f(M) / \partial M$ is a matrix whose ijth element is given by $\partial f(M) / \partial m_{ij}$.

We can state optimality conditions for the feedback gain Γ ($u = \Gamma x$) directly in matrix form (Athans, 1968)

$$\dot{X}^* = \partial H / \partial P^* = F(X^*, U^*)$$
$$\dot{P}^* = - \partial H / \partial X^* = - \partial L\,(t, X^*, U^*) / \partial X^* - \partial\, \text{trace}\, F(X^*, U^*)\, P^{*'} / \partial X^*$$
$$0 = \partial H / \partial \Gamma^*$$

To apply this result we need to compute gradient matrices. A partial list of useful expressions is given here, compiled from (Athans, 1968, Berger, 1976, see also Brewer, 1977)

$$\partial\, \text{trace}\, M / \partial M = I$$
$$\partial\, \text{trace}\, AMB / \partial M = A'B'$$
$$\partial\, \text{trace}\, MM' / \partial M = 2M$$
$$\partial\, \text{trace}\, AMBM / \partial M = A'M'B' + B'M'A'$$

where M is a matrix with independent entries. A typical application of the matrix maximum principle is the constrained L Q regulator. For linear dynamics

$$\dot{x}_t = Ax_t + Bu_t$$

and quadratic cost

$$J = \int_{t_0}^{t_f} (x_t' Q_1 x_t + u_t' Q_2 u_t) \, dt$$

it is desired to find the optimal linear feedback gain Γ ($u = \Gamma x$). With $X = xx'$ we get the matrix form of the dynamics under the proposed control

$$\dot{X}_t = (A + B\Gamma) X_t + X_t (A + B\Gamma)'$$

and the cost

$$J = \int_{t_0}^{t_f} \text{trace} \, [(Q_1 + \Gamma' Q_2 \Gamma) X] \, dt$$

Applying the above conditions yields

$$\dot{X}^* = \partial H / \partial P^* = (A + B\Gamma^*) X^* + X^* (A + B\Gamma^*)'$$
$$\dot{P}^* = - \partial H / \partial X^* = - Q_1 - \Gamma^{*\prime} Q_2 \Gamma^* - (A + B\Gamma^*)' P^* - P^* (A + B\Gamma^*)$$
$$0 = \partial H / \partial \Gamma^* = (Q_2 \Gamma^* + B'P^*) X^*$$

From the last equation we deduce

$$\Gamma^* = -Q_2^{-1} B' P^*$$

which, substituted in the optimal co-state equation, produces the familiar Riccati equation

$$- \dot{P}^* = A'P^* + P^*A - P^*BQ_2^{-1} B' P^* + Q_1 \qquad \text{with } P^*(t_f) = 0$$

Through matrix manipulations we have thus been able to derive the classical LQ regulator as a constrained optimization problem. The matrix maximum principle also provides convenient solutions to other constrained designs, like for example the LQ output feedback problem (Levine and Athans, 1970).

References

Anderson, B.D.O., and Moore, J.B. (1971). Linear Optimal Control, Prentice-Hall, New York.

Arnold, L., Crauel, H., and Wihstutz, V. (1983). Stabilization of linear systems by noise, SIAM J. Control Optim., 21 : 451.

Arnold, L.,and Wihstutz, V., Eds. (1986). Liapunov Exponents, Proc. Workshop Bremen University, Lecture Notes in Mathematics, vol. 1186, Springer Verlag, Berlin.

Aström, K.J. (1970). Introduction to Stochastic Control Theory, Academic Press, New York.

Athans, M. (1968). The matrix maximum principle, Inform. Control, 11 : 592.

Batkov, A.M. (1974). Optimisation Method in Statistical Control Problems, Mashinostroenie Moscow.

Bellman, R. (1957). Dynamic Programming, Princeton University Press, Princeton.

Bensoussan, A., and Lions, J.L. (1975). Nouvelles méthodes en contrôle impulsionnel, J. Appl. Math. Optim., 1 : 289.

Berger, C.S. (1976). The derivative of useful functions in control theory, Int. J. Control, 24 : 431.

Bertram, J.E. , and Sarachik, P.E. (1959). Stability of circuits with randomly time varying parameters, Trans. IRE, PGIT-5 : 260.

Brewer, J.W. (1977). The gradient with respect to a symetric matrix, IEEE Trans. Aut. Control, AC-22 : 265.

Bukhalev, V.A. (1980). Designing a control for markov process with random structure, Aut. Remote Control, 8 : 1141.

Caines, P.E., and Chen, H.F. (1985). Optimal adaptive LQG control for systems with finite state process parameters, IEEE Trans. Aut. Control, AC-30 : 185.

Elliott, R.J. (1982). Stochastic Calculus and Applications, Springer Verlag, New York.

Fleming, W.H. (1968). Optimal control of partially observable diffusions, SIAM J. Control Optim., 6 : 194.

Fleming, W.H., and Rishel, R.W. (1975). Deterministic and Stochastic Optimal Control, Springer Verlag, New York.

Has'minskii, R.Z. (1980). Stochastic Stability of Differential Equations, Sijthoff and Noordhoff, Alphen aan den Rijn.

Hijab, O.B. (1983). The adaptive LQG problem, IEEE Trans. Aut. Control, AC-28 : 171.

Joseph, P.H., and Tou, J.T. (1961). On linear control theory, Trans. AIEE Appl. Industry, 80 : 193.

Kalman, R.E., and Bertram, J.E. (1960). Control system analysis and design via the second method of Liapunov, Trans. ASME, J. Basic Eng., 82 : 371.

Kats, I.I., and Krasovskii, N.N. (1960). On the stability of systems with random attributes, J. Appl. Math. Mech., 24 : 1225.

Kazakov, I.E., Statistical Theory of Control Systems in State Space, Fizmatgiz, Moscow.

Kozin, F. (1969). A survey of stability of stochastic systems, Automatica, 5 : 95.

Kushner, H.J. (1965a). On the stochastic maximum principle : fixed time of control, J. Math. Anal. Appl., 11 : 78.

Kushner, H.J. (1965b). On the stochastic maximum principle with "average" constraints, J. Math. Anal. Appl., 12 : 13.

Kushner, H.J. (1967). Stochastic Stability and Control, Academic Press, New York.

Levine, W.S., and Athans, M. (1970). On the determination of the optimal output feedback gains for linear multivariable systems, IEEE Trans. Aut. Control, AC-15 : 44.

Pardoux, E., and Talay, D. (1985). Discretization and simulation of SDE, Acta Applicandae Mathematicae, 3 : 23.

Pontryagin, L.S., Boltyanskii, V.G., Gamkrelidze, R.V., and Mishchenko, E.F. (1962). The Mathematical Theory of Optimal Processes, Interscience, New York (translation).

Rishel, R.W. (1981). A comment on a dual control problem, IEEE Trans. Aut. Control, AC-26 : 606.

Simon, H.A. (1956). Dynamic programming under uncertainty with a quadratic criterion function, Econometrica, 24 : 74.

Sworder, D.D. (1968). On the stochastic maximum principle, J. Math. Anal. Appl., 24 : 627.

Sworder, D.D. (1969). Feedback control of a class of linear systems with jump parameters, IEEE Trans. Aut. Control, AC-14 : 9.

Varaiya, P. (1975). The martingale theory of jump processes, IEEE Trans. Aut. Control, AC-20 : 34.

Warfield, V.M. (1976). A stochastic maximum principle, SIAM J. Control Optim., 14 : 803.

Witsenhausen, H.S. (1968). A counterexample in stochastic optimum control, SIAM J. Control Optim., 6 : 131.

Wonham, W.M. (1968). On the separation theorem of stochastic control, SIAM J. Control Optim., 6 : 312.

Index